Rhetoric and Ethics in the (

It has become increasingly difficult to ignore the ways that the centrality of new media and technologies—from the global networking of information systems and social media to new possibilities for altering human genetics—seem to make obsolete our traditional ways of thinking about ethics and persuasive communication inherited from earlier humanist paradigms. This book argues that rather than devoting our critical energies towards critiquing humanist touchstones, we should instead examine the ways in which media and technologies have always worked as crucial cultural forces in shaping ethics and rhetoric. Pruchnic combines this historical itinerary with critical interrogations of diverse cultural and technological sites—the logic of video games and artificial intelligence, the ethics of life extension in contemporary medicine, the transition to computer-automated trading in world stock markets, the state of critical theory in the contemporary humanities—along with innovative analyses of the works of such figures as the Greek Sophists, Kenneth Burke, Martin Heidegger, Michel Foucault, Friedrich Nietzsche, and Gilles Deleuze. This book argues that our best strategies for crafting persuasive communication and producing ethical relations between individuals will be those that creatively replicate and appropriate, rather than resist, the logics of dominant forms of media and technology.

Jeff Pruchnic is an Assistant Professor in English at Wayne State University, USA.

Routledge Studies in Rhetoric and Communication

Rhetoric and Ethics in the Cybernetic Age

The Transhuman Condition

Jeff Pruchnic

Routledge
Taylor & Francis Group

NEW YORK AND LONDON

First published 2014
by Routledge
711 Third Avenue, New York, NY 10017

Simultaneously published in the UK
by Routledge
2 Park Square, Milton Park, Abingdon, Oxfordshire OX14 4RN

First issued in paperback 2016

*Routledge is an imprint of the Taylor & Francis Group,
an informa business*

Library of Congress Cataloging-in-Publication Data
Pruchnic, Jeff.
Rhetoric and ethics in the cybernetic age : the transhuman condition / by Jeff Pruchnic.
p. cm. -- (Routledge studies in rhetoric and communication ; 17)
Includes bibliographical references and index.
 1. Internet—Social aspects. 2. Internet—Moral and ethical aspects.
 3. Information technology—Social aspects. 4. Information technology—Moral and ethical aspects. I. Title.
  HM851.P768 2013
  302.23'1—dc23
  2013011250

ISBN 13: 978-1-138-21874-1 (pbk)
ISBN 13: 978-0-415-84034-7 (hbk)

Typeset in Sabon
by IBT Global.

For my wife, Millie Rovan Pruchnic, and my daughter, Stella Grace Pruchnic, who inspire me every day.

Contents

Figures

Acknowledgments

In *The Genealogy of Morals*, Nietzsche, that most prescient and pre-eminent thinker of the transhuman, identifies "debt," particularly in the form of personal obligations, as a quintessentially human quality, and one that has long been the source of many of the species's more destructive tendencies and stood in the way of human "self-overcoming." Insofar as this book takes a number of its most important cues from Nietzsche's thought, this may put me in an awkward position—because I owe a number of debts to a wide variety of individuals who both inspired and supported its writing. However, given that one of this book's central arguments is for the potential of productively redirecting many of the negative qualities associated with "human nature," perhaps we can just consider this acknowledgments section an example of that strategy.

The research for this book began as a dissertation project in the English Department of the Pennsylvania State University, and I owe perhaps my largest debt to my dissertation director and mentor, Richard Doyle ("To Doc, with Love"), and the other members of my committee: Stephen H. Browne, Cheryl Glenn, Jeffrey T. Nealon, and Jack Selzer. An even more long-standing debt is owed to Joseph O. Dewey, teacher and friend for almost two decades and counting. I am similarly grateful to my friend and frequent collaborator Antonio Ceraso; this book would have been far poorer without the insight and advice he has provided over the years.

This book also owes much to the immense support I have received from Wayne State University and its Department of English. Ellen Barton, Gwen Gorzelsky, and Richard Marback have been remarkable friends and mentors over the past several years, as have been the extraordinary group of scholars and teachers that make up Wayne State's English Department as a whole. The writing of this book has also benefited from a significant amount of institutional support from Wayne State, including a semester sabbatical, a summer research grant, and financial support of archival research. Much of the work between these pages also found its first form as talks delivered as part of the Wayne State Humanities Center's colloquium series, overseen by the always remarkable Walter Edwards.

I also owe a debt of gratitude to a large number of other individuals who offered suggestions, advice, or a kind word during the completion of this project: Robert Aguirre, Dana Anderson, David Blakesley, James J. Brown, William Covino, Rosa Eberly, Jamie Ebersole, Jessica Enoch, Theresa J. Enos, Alexander R. Galloway, Richard Grusin, Kathryn Hume, Jay Jordan, Kim Lacey, Carolyn R. Miller, Jodie Nicotra, Jeff Rice, Jenny Rice, Scott C. Richmond, Mike Ristich, Brian Rotman, Stephen Schneider, Sanford Schwartz, Michael Scrivener, Stuart Selber, Steven Shaviro, Paul Dustin Stegner, Allan Stoekl, Barrett Watten, Scott Wible, Elizabeth S. Wilson, Lynn Worsham, Anne Frances Wysocki, and James T. Zebroski.

I was also very lucky to have Elizabeth Levine and Emily Ross (Routledge) as my editors on this project, who expertly guided this manuscript from submission to publication. On the production side, I am also thankful to Anke Reisenkamp and the Van Gogh Museum, and Hannah Rhadigan and the Artists Rights Society, for help clearing permissions for visual material. An earlier form of Chapter 3 of this text, "Rhetoric in the Age of Intelligent Machines," was published in 2006 by *Rhetoric Review* under the title "Rhetoric, Cybernetics, and the Work of the Body in Burke's Body of Work"; I am grateful to the editors of *Rhetoric Review* for permission to republish that material.

Last, but certainly not least, I owe very much to my parents, Albert and Sharon Pruchnic, and my sister, Jennifer Pruchnic, for being my first and most important teachers in what it means to be human.

Introduction
The Cybernetic Age

One can of course see how each kind of society corresponds to a particular kind of machine—with simple mechanical machines corresponding to sovereign societies, thermodynamic machines to disciplinary
societies, cybernetic machines and computers to control societies. But
the machines don't explain anything, you have to analyze the collective
arrangements of which the machines are just one component.

> —Gilles Deleuze, "Control and Becoming" (175)

The thought of every age is reflected in its technique.

> —Norbert Wiener, *Cybernetics; or, Control and Communication in
> the Animal and Machine* (38)

This book is about the impact of contemporary technology and new forms
of media on cultural life in the present. In this sense it joins a wide variety
of other analyses that have drawn our attention to how computing technologies became a pivotal factor in human experience over the past several decades, not only altering forms of communication and the shape of
various industries, but also having a more general effect on social life in its
entirety, on politics both local and global, and on "nature" as both our ecological environment as well as what we take to be "natural." However, the
central arguments of this book cut against what largely remains the default
conception, or critique, of such phenomena within both cultural theory
of the humanities and social sciences as well within popular culture as a
whole. Rather than presuming that increased centrality of technology and
media within culture has led to an increased standardization, "dehuman-
ization," or homogenization of culture and human experience, this project
argues instead that the fundamental logic of contemporary technology and
media has been one of an increased "humanization" of social and technical
systems of diverse types, one in which increased flexibility, the mimicking
of biological systems of feedback and adaptation in mechanical realms, and
the microtargeting of the dispositions and desires of increasingly smaller
groups of people have become the key drivers of economic and cultural
production in the present.

Thus, for instance, advances in flexible specialization and niche marketing make it possible to develop and promote products in response to ever-smaller changes in consumer preference and in pursuit of ever-smaller markets (approaching the servicing of, as announced in the title of a recent popular advertising text, "a market of one"). The arrangement of political campaigning around appeals to proponents of particular parties or identity categories (the "soccer moms" and "Nascar dads" that we heard so much about in the 1990s) are radically refined, in accordance with advanced data-mining and aggregation techniques, into targeted groups based on the cross-indexing of a near-infinity of consumer preferences and "social interest issues." Bioinformatics research in the life sciences, once a key cultural symbol of the acontextual and potentially dangerously reductionist tendencies of contemporary epistemologies and representational schemas, are transformed, in the "postgenomic" age, toward the production of specialized treatments and preventive procedures based on individual genetic profiles. Educational methods previously dedicated to the transmission of standardized content give way to "adaptive learning environments" in which discrete skills are taught in accordance with a students' particular learning styles and problem areas. In aesthetics, artworks increasingly rely on the active participation of their audiences, and design methods in architecture and manufacturing increasingly make use of "algorithmic" methods for imitating organic forms and "evolutionary" formulas for maintaining structural integrity. The list inexorably goes on.

I argue in this book that such phenomena are the results of a broader shift in the fundamental logic subtending contemporary culture over the past several decades. More specifically, I suggest that such changes are hard for us to directly thematize or account for because they create a fairly radical rearrangement and overlap of two domains that were separated within Western culture nearly six centuries ago: *logos* (the realm of communication, reason, representation, and the "natural" most generally) and *techne* (techniques and material technologies that imitate or take the place of natural phenomena). One of the central theses of this book is that this separation has been a remarkably consistent foundation for our understandings of ethics and politics for several centuries, providing the basis for discriminating between "ends" and "means" in moral thought, as well as, in regard to questions of political agency and social transformation, the status of such categories as power and resistance. However, my larger argument is that our reliance on this division and these categories is very much at the heart of a number of a number of ethical and political dilemmas prominent in contemporary culture—from the difficulties of deciding on boundaries for the manipulation of human genetics, to how we might imagine autonomy in a world of increased social and physical connection—and that we may have to think through and beyond such commonplaces in order to develop effective responses to the ways that culture has been altered in the half-century or so during which computing technologies and new media have become increasingly central to social life.

In Chapter 1 of this book, I present a much more detailed periodization of contemporary culture around these changes and their specific relationships to pressing ethical and political questions of the present. Here, in this introduction, however, it is perhaps useful to begin with some crucial background on a few of these contexts and some key terms highlighted in this text's title and pages, as well as to account for why I suggest rhetoric as our primary disciplinary domain and field of praxis for this endeavor.

THE CYBERNETIC AGE

Surviving Greek texts of the late third and early fourth century BCE record the stabilization of two terms that had previously held many indistinctions, *logos* (λόγος, our root for "logic") and *techne* (τέχνη, our root for "technical"). *Logos* would come to denote not only human reason and rationality, but discourse, calculation, and the principles governing matter, energy, and the terrestrial environment, phenomena bound together by their associations with the "true" and the natural. *Techne*, conversely, would be formalized to refer to technical crafts and aesthetics, as well as cunning, wile, and deception, qualities joined by their "false" or artificial character. In addition to signaling, as numerous classicists have suggested, the end of the significant influence of *mythos* or mythological thinking as a structuring principle of Greek cultural and intellectual life, the definition of *logos* over and against *techne* is repeatedly invoked in Classical Antiquity to found a large number of more specific divisions and associations, categories that continue to cast a long shadow over the present: proto-humanist conceptions of human examplarity based on its distance from irrational animals and lifeless matter, beginning with the ontological positioning of human beings between those of "brute" animals and the divine; the classification of distinct modes of observation and analysis, a sowing of seeds for the increasingly specific domains of the sciences and humanities; the forwarding of the human *psyche* as, alternately, a "mind" or "soul" that is distinct from, or at least not reducible to, the body; the formal emergence and disidentification of philosophy and rhetoric, two fields competing for recognition as the disciplinary domain for the study and dissemination of wisdom; and the ethical separation between moral and material (that is to say, economic) "value," and the alliance of ethical ends with ethical means. Given all of these developments, it might not be hyperbolic to suggest this division between *logos* and *techne* was one of the most pivotal within the entire history of Western intellectual thought.

It was not always this way. Indeed, in tracing the intersections of rhetoric, technology, and politics in the present moment, this book also in many ways follows a lexigraphic and conceptual overlap of these domains best captured by a concept coeval with the separation of *logos* and *techne*, and one that would return to prominence in the mid-twentieth century.

The ancient Greek term *kybernetes* (κυβερνήτης) was often used in the fourth century BCE to denote objects (such as the rudder or a ship) or individuals (such as a steersman or pilot) that directed, but did not fully control, some other object or system, often through artificially simulating a natural process or force. *Kybernetes* would later be the basis for the Latin *guberno*, and thus our root for "governance." In 1843, the French physicist Andre-Marie Ampere would recover much of the term's original meaning in using *cybernetique* to describe those operations of government that relied more on manipulating the structures of political economy than on the direct exercise of sovereign power.[1] In Plato's *Gorgias*, *kybernetes* is used to define (and indeed coin the proper name for) rhetoric (*rhetorike*), the study and practice of persuasion (511d–513c). As it often was at the time, in this dialogue *kybernetes* and rhetoric are both associated with *techne*, as well as with *mechane* (artifice and strategic deception), our root for "machine." For instance, Plato's student Aristotle would later define the "mechanical" and the "rhetorical" as forces that mutually disrupt or invert natural processes in his *Rhetoric* and *Mechanics*. In the former, the rhetoric of the sophist Corax is condemned for subverting the natural order of things by "making the worse argument seem the better" (1402a24–25); in the opening pages of the latter, "mechanical skill" is identified as the force that allows us to act "contrary to nature" by creating a situation in which the "less prevails over the greater," such as in using a lever to move a large weight (847a10–25). However, in the work of Plato, the similarities between *techne* or *mechane* and rhetoric would be used to divide all three from another category. In comparing rhetoric and *kybernetes*, Socrates' condemns the rhetoric practiced by the Greek sophists as an instrumental *techne* inferior to the pursuit of "true knowledge" (*episteme*) carried out by philosophy, thus setting the future path for Western metaphysics and reason.

Centuries later, in the early 1940s, the American mathematician Norbert Wiener would appropriate this reference to *kybernetes* in the *Gorgias* as the title for an emergent interdisciplinary inquiry into "control and communication in the animal and the machine," an endeavor he would later gloss as encompassing "not only the study of language but the study of messages as a means of controlling machinery and society, the development of computing machines and other such automata, certain reflections upon psychology and the nervous system, and a tentative new theory of scientific method" (*Human* 15). Wiener's coining of "cybernetics" may be an accident of history; he had originally wanted to modify *angelos*, Greek for "messenger," but did not want the unavoidable association with "angel" (*Mathematician* 322). If so, however, it was a fortunate one, as the work that went on under the banner of cybernetics in the mid-century may be taken as paradigmatic for not only the subsequent development of computing technologies, biotechnology, and digital media, but the emergence of *techne*'s priority in structuring political, economic, and cultural life today.

The discipline that would later become known as cybernetics formed in a series of conferences sponsored by Josiah Macy Jr. Foundation between 1946 and 1953.[2] One of the organizing objectives of the conferences was to encourage discussion between representatives from fields that might not normally be in collaboration; as such, the groups assembled in the ten Macy Conferences, none of which had more than two dozen participants, drew members from such disparate disciplines as engineering, biology, psychology, neurophysiology, sociology, zoology, mathematics, mathematics, anthropology, pharmacology, philosophy, and ethnology. Conference participants that were or would later become luminaries in their respective fields included Norbert Wiener himself, whose wartime research would prefigure the key concepts of the field and who would popularize the movement for non-specialist audiences in such texts as *Cybernetics* (1948) and *The Human Use of Human Beings* (1950); Wiener's frequent early collaborators, the Chilean physiologist and physician Arturo Rosenblueth and the electrical engineer Julian Bigelow; the polymath scientist Jon Von Neumann, a principal member of the Manhattan Project who would subsequently pioneer game theory and make important contributions to computer architecture; the neurologist Warren McCulloch, who, in collaboration with the enigmatic prodigy and autodidact Walter Pitts, would produce landmark work in neural network theory; Kurt Lewin, often regarded as the founder of social psychology; J. C. R. Licklider, a computer scientist who made crucial contributions to the Cold War "Semi Automatic Ground Environment" computer-aided defense system and the Internet predecessor ARPANET; Paul Lazarfeld, one of the first prominent sociologists to address the impact of mass media and the founder of Columbia University's Bureau for Applied Social Research; and the anthropologists Margaret Mead and Gregory Bateson, pivotal figures in the study of (respectively) human sexuality and systems theory and ecology. As one might expect given the disparate disciplines and objects of research (human bodies, social and environmental systems, animals, machines) represented by members of the cybernetics group, their collective work coalesced around a number of principles that cut across these various domains. While there are certainly interesting histories to be written about the legacy of any number of these principles within particularly disciplinary fields, my concern here is in outlining two in particular that I will suggest have transformed over the past few decades from being specialized ways of orienting scientific research to become, through their integration into a variety of new technologies and media, general structuring principles of contemporary culture, communication, and forms of social power.

First and foremost was a radical reconsideration of the *ecological* interaction between actor and environment seen most clearly in cybernetics' recurrent interest in processes of feedback, equilibrium, and homeostasis. Rosenblueth had been a collaborator of physiologist Walter Cannon, who coined the term "homeostasis" to designate the process through which an organism's internal environment maintained a stable condition in the face of a changing external environment. In extending the mid-nineteenth century work of Claude Bernard, the French physiologist who first hypothesized an "interior environment" [*milieu intérieur*] regulating the fluid environments inside of organisms, Cannon both provided a more thorough explanation of this process in multicellular life and, in a deliberately provocative epilogue to his 1932 book *The Wisdom of the Body*, suggested that the regulatory principles of homeostasis might be extended to study the cooperation of individuals within civil society. The work of the cybernetics movement would take up the organic and the social applications of homeostasis but also argue for its relevance in mechanical processes and, perhaps most importantly, for studying the interactions between humans and machines in complex systems. In his wartime research for the U.S. government, Wiener had studied how pilot reaction times, aircraft speeds, and the firing mechanisms of antiaircraft weaponry could be calculated as a single system in designing best practices for aerial combat; the cybernetics movement would expand this approach to both propose more complicated symbiotic or interactive systems and to suggest ways that human physiology and behavior might itself be viewed as evincing qualities more commonly associated with machines. Although the legacy of cybernetics in the popular imagination—perhaps because of the coining of the term "cyborg" for "cybernetic organism" by Manfred Klynes in 1960—became associated with research into the material combination of human bodies and machinery, the focus of the movement and their subsequent impact on the various fields affected by cybernetics research would be more correctly described as the overlapping of "natural" and "artificial" (or biological and mechanical) *processes* and *systems*, one that began with the conceptual extension of a organic process (homeostasis) into inorganic realms and then fed backward into a focus on how the increasing complexity of mechanical and technological processes—and their increasing impact on the biological and social—showed these latter two realms to have a logic not altogether different from the technical.

Such a systems-oriented perspective also required a rethinking of the *teleological* as a concept in the sciences as well as the teleology of science itself: the objectives or ends driving scientific praxis. Indeed, a reconceptualization of "ends" was in many ways the starting point of what would later be known as cybernetics: a series of papers by Wiener, Rosenblueth, and Julian Bigelow on the concepts of behavior, purpose, and teleology, largely inaugurated the interdisciplinary projects that prompted the Macy gatherings and the development of cybernetics as a recognizable field. In these writings, early members of the cybernetics group argued that assigning "purpose" as an attribute available only to humans was quickly becoming a moribund distinction in the face of technologies designed with defined goals and the ability to flexibly respond to their environment. Here too, the social and technological would meet somewhere in the middle. On the one hand, the agency of purpose was distributed into the technological domain: "if the notion of purpose is applicable to living organisms, it is also applicable to non-living entities when they show the same observable traits of behavior" (Rosenblueth and Wiener 325); on the other, however, the study of how machines might be designed to respond flexibly to changing circumstances while maintaining fidelity to their "purpose" also suggested how one might use the same process to predict the strategic behavior of humans within particular scenarios, an objective von Neumann would later pursue in his pioneering work in game theory.

Redefining purpose as a force that could operate outside of human agency and without a determinate goal other than the maintenance of a system's overall equilibrium was not only immensely salutary for the cyberneticists' design of "control systems" for automata and machinery, but also for assaying the ways biological and social systems could be robust in form but adaptable over time. In addition to von Neumann's application of game theory to political decision making and international relations, McCulloch and Pitts would hypothesize a propositional logic for analyzing how the brain formed memories and how an individual's previous experience influenced their future actions, and Pitts and Wiener would separately theorize how emotional response and affective states were similarly attuned by the conditioning of the nervous system. However, redefining teleology in such a manner was also aligned with an emerging sea change in the purpose of science itself; as Weiner observed, cybernetics was at the forefront of a larger process through which scientific research became less about understanding global questions of how the natural world functioned and more about the strategic investigation of natural, organic, and mechanical processes in the pursuit of particular goals, and thus a breakdown between the control sciences (such as engineering) that were more traditionally associated with such methods and the natural or life sciences. Although the latter, since at least the time of Aristotle, were traditionally driven by the desire to understand the natural state of systems and restricted to therapeutic interventions to restore this state, from mid-century onward they would increasingly become organized around the deliberate manipulation of genetic material and physiological response. Although Wiener emphasized how in the wake of cybernetics "the vocabulary of engineers soon became contaminated with that of the neurophysiologist and psychologist," the opposite flow of influence has certainly became equally if not more important to the function of scientific praxis today, one written in the very names of such newer fields as bioinformatics, biotechnology, and bioengineering (*Cybernetics* 15).

Cybernetics enjoyed international prominence in both scientific circles and the popular consciousness well into the early 1970s. In the U.S. it was embraced by numerous American academics, many of which promoted it as a *scienza nuova* with universal application to diverse fields of research. In mid-century Soviet science and politics, cybernetics moved from being viewed as an imperialist pseudoscience in the 1940s to being embraced by the early '60s as an approach to science, technology, and sociology particularly well-suited to communist ends. Under the auspices of Salvador Allende, cybernetics management theorist Stafford Beer oversaw the "Cybersin" project of creating a computer-automated "real-time" planned economy in early 1970s Chile (remarkably useful during the Chilean truck driver's strike of October 1972, its control center was destroyed in the coup of the following year).[3] In European philosophical thought, Martin Heidegger famously considered whether the advent of cybernetics marked the end of Western metaphysics (a point I return to in Chapter 2), Jacques Lacan devoted an entire seminar to the question of what psychoanalysis might learn from cybernetics, and in his early influential work *Of Grammatology*, Jacques Derrida would position cybernetics as a conceptual movement that overlapped with his own deconstructionist technique (though one that ultimately fell short in failing to interrogate its own "historico-metaphysical character") (9).[4]

By the late 1970s, however, cybernetics would largely disappear as a recognizable discipline or methodological approach within the sciences. This denigration was, at least partially, due to the inability of the cybernetics movement to achieve Weiner's goal of fighting against the grain of the ever-deepening specialization of scientific research. In this sense, cybernetics was in many ways undone by its own success; demonstrating the centrality of feedback systems and networked interaction in various domains suggested the potential of sub-disciplinary foci on these particular sites, thus paving the ways for disciplines specializing in the intersection of previously autonomous areas of study: computer science, neuropharmacology, molecular genetics, ecology.

In an interview near the end of his life, Heinz Von Foerster, one of the youngest members of the cybernetics group and the official archivist of the relevant Macy Conferences, was asked to reflect on the decline of cybernetics and its legacy the end of the twentieth century. When queried about why movement never went "mainstream," Von Foerster suggest that cybernetics has not so much disappeared but disseminated so widely across contemporary culture, science, and technology as to appear invisible (Waters 81). Although cybernetics may be dead as a formalized discipline, Von Foerster suggests that "underneath it's completely alive"; indeed, he goes on to suggest that cybernetics is all the more powerful and influential today for not being explicitly recognized or expressed in all the fields it has intersected, making it both harder to reject and easier for it to function as what he calls an "underground" connecting disparate vectors of scientific praxis and cultural experience. Although cybernetics may not have become mainstream over the past half-century, the mainstream of science, technology, and society had itself become cybernetic.

One of my more overarching arguments might be taken as a radicalization of Von Foerster's suggestion here: a central thesis to which I continually return throughout these pages is that the most pervasive impact of the age of information technology after cybernetics is not the increasing ubiquity of these material technologies themselves, but the ways in which politics, culture, and economics has increasingly found its operating principles in those processes that find only their most obvious manifestation in physical technology—*techniques*. In other words, if there is a singular logic that can be used to map the changes in social power that have occurred over the last few decades, the suture point connecting contemporary cultural production, economics, and political and social discourse, it would be what we might call, for lack of a better term, a *technologic*: forms of interaction and engagement that not only find their most explicit manifestation in contemporary technologies but signal the imbrication, or recombination, of *techne* (formalistic and goal-directed strategies) with *logos* (both in its sense of human "higher reason" and of the general structuring of human life) that Plato attempted to so carefully separate and the division of which became a touchstone for Western thought.

Another recurrent argument in this book is that such changes make rhetoric a particularly privileged domain for analyzing contemporary culture and for devising concrete ethical and political strategies within this environment. Although, as mentioned above, Wiener likely did not have rhetoric specifically in mind when he chose the name cybernetics for his prototypical inquiry into information technology and the study of social systems, the two perspectives are remarkably similar in a number of ways. Indeed, as prominent theoretical physicist Satosi Watanabe, a key thinker in both quantum mechanics and what would become known as "second order cybernetics" research, Wiener's selection of the term from its context in the *Gorgias* was prescient for cybernetics' relationship with traditional scientific epistemological frameworks:

> [I]t is highly significant that in his mind Plato somehow associated rhetorics and cybernetics. We should notice that these two arts have indeed something in common: They both represent flexible and adaptive methods aiming at utilizing, influencing, controlling, and overcoming the outside world, mental or physical, in order to achieve one's own goal. They are entirely different from primarily disinterested sciences such as geometry or astronomy or from straight technology such as bridge-building or oil pressing. (152)

One of the primary objectives of this book is to analyze how this shared domain of cybernetics and rhetoric has become the dominant logic of contemporary culture. In Chapter 2, I present a brief genealogy of the dominant modes of combining technics and media over several centuries, leading up to what I call the "parametric" present, in which forms of knowledge and representation are not longer primarily based on quantification (measurement) or calculation (prediction and relation), but dynamic processes that work to maintain relationships between a variety of elements. Concomitantly, whatever we might call social conditioning or social power in this regime works primarily through logics of inclusion rather than exclusion, an attempt to integrate and draw value from as many heterogeneous identities or behaviors as possible. Chapter 3 revisits the early work of key rhetorical theorist Kenneth Burke and his own responses to prototypical developments in cybernetics and information technology to argue for how we might account for the ways such developments have fundamentally changed human persuasion, and in particular the role of human affective or precognitive dispositions, in rhetorical processes. Chapter 4 continues this focus on the intersections of rhetoric, information technology, and new media in an effort to rethink and extend Gilles Deleuze's writings on the "control societies" of the present. Finally, in Chapter 5, I return in many ways to the "primal scene" of both the separation of *techne* and *logos* as well as of rhetoric and philosophy to argue for how we might rethink ethics in an age of global media, biotechnology, and hypercapitalism.

COMING TO TERMS:
TRANSHUMANISM, RHETORIC, ETHICS

The other major inspiration for this book, and the source of the term transhuman, comes from the writings of the prominent mid-century biologist Julian Huxley. Huxley is perhaps best known today for his contributions to what became known as "the modern evolutionary synthesis," a prominent paradigm in evolutionary biology, and his role in the founding of the United Nations Educational, Scientific, and Cultural Organization (UNESCO), as well as for being the grandson of Darwin's notable contemporary and champion, Thomas Henry Huxley, and the brother of the famous Aldous Huxley. However, from the 1930s onward, Huxley also produced a large and wide-ranging series of writings on the political import of scientific discoveries as well as, in continuation of an interest that occupied his famous grandfather, the more specific question of the role of evolution on historical and contemporary forms of social behavior and ethics.

In the short but prescient essay "Transhumanism," first published in 1957, Huxley predicted that our abilities to better understand and replicate processes of the natural world, as well as to more directly and more significantly intervene into natural processes, would soon lead to a much greater cross-coupling of, or growing indiscernability between, the natural and the artificial as conceptual categories, and that this would concomitantly require something of a sea change in human ethics, or at least a significant alteration in the methods, if not the goals, of humanist ethics. Huxley presents two distinct meanings for what he calls an emergent "transhumanism" that might accomplish such an objective. First, Huxley predicts that the energy typically devoted to the scientific exploration of the world and the creation of material technologies is crossing over into the development of techniques for the direct manipulation of the natural environment and the creation of social processes (educational, economic, political) that themselves replicate the complex forms found in natural ecologies and reproduced in mechanical ones. As Huxley writes, now that humans have "pretty well finished the geographical exploration of the earth" and have "pushed the scientific exploration of nature, both lifeless and living, to a point at which it main outlines have become clear" their next step is the integration of such findings into social processes of various types (14). Such a situation also created for humans, and for the doctrine of humanism, what Huxley will refer to as an inevitable ethical burden in regards to the protection of the natural world and the administration of justice within social formations. The very ability of humans to massively alter the natural world, and the withdrawal of recognizable theological and indigenous traditions of ethics, made it impossible for humans to defer to categories of "the natural" or to transcendental values of other types as unquestionable resources for ethical reason and decision making. As Huxley refers to it, the human species has been "appointed managing director" of the "business of evolution," in both the terrestrial and social environment (13).

Although the term transhumanism hast been used for different purposes in more recent times (of which, more in the next chapter), my use of the term here is heavily indebted to Huxley's thought in this regard. Broadly speaking, Huxley's thesis about the ways in which artificial and social processes would take on the ecologically complex and equilibrium-based logic traditionally aligned with evolution and other biological and natural processes, very much informs my approach to thematizing cultural change after cybernetics. His argument for the unavoidable ethical responsibilities of individuals in a time humans have largely taken over the design and maintenance of such processes "from" nature, on the other hand, provides, as we shall also see in more detail in the next chapter, a framework for analyzing the role and function of rhetoric and ethics in the present.

However, I also expand on Huxley's concept here in reference to four other contemporary phenomena that were absent or only implicit in Huxley's work. Conveniently, for my purposes here, all of them might be arranged under additional terms sharing the prefix "trans." The first is the entry of the category of "the human" into a state of continuous *transition*. As several decades of critical work on performativity and social identity have confirmed, consensus of what "is" or can be accepted "as" human— or of what is constitutive of humans as opposed to non-human animals or machines—is a determination that has long undergone processes of negotiation, expansion, and contraction. Yet, over the past several decades, and in response to a variety of advances in our abilities to simulate human behavior in mechanical realms and alter human genetic material, this implicit process becomes an increasingly explicit and urgent concern for a variety of specialized communities—life scientists, philosophers, bioethicists, legislators—as well as within contemporary culture as a whole.

Second, I use the term *transference* to refer to the particular ways in which generic capacities of humans, non-human animals, and machines become shared or cross-coupled across these domains, a process that occurs between the invention of particular information technologies and biotechnologies and their integration in human praxis and experience. Although the lines drawn between the human, the non-human animal, and the machine are often thought of in terms of identity or ontology—whether such a hybridity or overlap is positively celebrated by Donna Haraway in her paradigmatic presentation of the cyborg as a "political fiction" that eschews the rigid classifications of history in favor of "permanently partial identities" (154), or decried, as in the work of political philosopher Francis Fukuyama, as a dangerous encroachment on an "empirically grounded view of human specificity" that is foundational for liberal democracy (*Posthuman* 147)—these considerations are themselves largely posterior to the transference of animal and machinic capacities. Thus, the pursuit of artificial intelligence within computing technologies is an attempt to simulate the capacities associated with human thought, and the success or failure of artificial life is evaluated on whether it demonstrates the capacities we associate with organic life (mobility, differentiation, reproduction, etc.). On the opposite end of the scale, biotechnology, as its name itself suggests, finds its vocation in creating ways in which the organic manner of humans and other animals can take on the combinatory and programmatic capacities we associate with machines and mechanical processes. While the transitional state of human identity or ontology has recently returned as a crucial issue in legal and ethical thought, particularly in relation to non-therapeutic manipulations of human genetics (questions I address in Chapters 1 and 5), throughout this book I also pursue the ways that the transferences of capacities has come to structure politics, culture, and communication in a broader sense.

Third, I use the term *transactional* to describe common processes in technology and political economy that might be taken as the actualization of what the cybernetics group could only theorize in the abstract: procedures that find their identity and coherence through the process of transformation itself or the pursuit of a particular equilibrium rather than a more discrete goal. The variety of computer coding that undergirds information technologies such as computer operation systems and their larger linkages (such as computer networks or the Internet) are perhaps the most visible examples of such transactional processes as well as the ones most indebted to early cybernetic work on this subject. Much like cyberneticists' interest in substitutional and translational grammars for describing and duplicating automated processes—such as McCulloch and Pitts' logical calculus and cyberneticists' broader fascination with "black box" thought experiments in which only inputs and outputs were registered—software coding has no "significance" (no deeper meaning to be revealed), but the lack of such significance is proportional to its overall flexibility for representing, simulating, and categorizing diverse phenomenon. In developing the transactual as a cultural tendency in this text, I am particularly inspired by Foucault's references in some of his later lectures to the "transactional realities" [*réalités de transaction*] that occur at the interfaces as social and technical processes, ones that are not natural in the strictest sense but also not "false" or open to being disproved (*Birth* 297). As opposed to the "virtual," a more common term for such phenomena in the context of technology, the transactual has the advantage of emphasizing not only the consideration of a phenomenon as the imitation of an actual process, but also the ways in which it is itself created by the coordination or combination of dynamic processes. I argue that such transactional process have become the dominant logic of a variety of cultural domains in Chapter 3 of this text, and in Chapter 5, I use the term transactional ethics as a descriptor for the question of whether it is possible to consider "value" in the sense of ethics or morals and in its purely economic sense as complimentary (or at least non-oppositional) processes.

Finally, I use the term *transversal* to refer most broadly to unusual connections between other domains or entities that we may have once kept separate (such as the natural and the artificial, or the human, the non-human animal, and the machine). The transversal is also used here in a strategic sense, one drawn from Félix Guattari's use of the term. In a 1964 essay of the same name, Guattari presents "transversality" [*transversalité*] as a counterpoint to the psychological concept of therapeutic transference. In the traditional understanding of transference, a patient's feelings for another individual become transferred to their psychotherapist; Guattari's forwarding of the transversal was, on the one hand, an attempt to think through how such transfers of ideational and affective investment might take place in ways that do not follow a traditional hierarchy (the patient subordinate to the therapist) or well-worn sequence, thus avoiding what he codes "the impasse of pure verticality" presumed in such interactions (18). On the other hand, however, Guattari was also equally convinced that the forming of such connections was unavoidable, and thus it is naive to believe that a "mere horizontality" of non-hierarchical relations was possible or that a "natural" or originary subjectivity or identity was waiting to be freed within individuals by their overcoming of extant attachments and relationships. While in later works Guattari would use the term transversal as a general reference to the variety of domains (politics, aesthetics, biology) that shape human identity, my own use of the transversal as a rhetorical strategy here hews back to Guattari's original interest in term to describe the ways that existing affective investments might be redirected in novel ways or connected to different objectives. In particular, I argue throughout this text that our most productive contemporary strategies for intervening in communal relations (ethics) and crafting motivational and persuasive tactics (rhetoric) may have to be premised on the creation of such transversal connections—ones that often work, to paraphrase T. S. Eliot, by motivating individuals to do "the right thing for the wrong reason."

My approach to ethics and rhetoric in this book might also be in need of some explanation, not because these categories have, as with transhumanism and cybernetics, short or obscure histories, but for the opposite reason: because they have extraordinarily long histories within Western intellectual thought and are often defined in contradictory ways. At the risk of overly schematizing one of these immensely long histories, we might posit that debates over the proper identity of rhetoric and its appropriate role in analyzing contemporary culture and communication largely take place around and between two contrasting positions; while these two different definitions have certainly taken on new exigencies and meanings in recent history, both might also be read as repetitions of a sort of conflicts in early Greek culture from which rhetoric initially emerged.

The first is the contention that rhetoric is largely a series of techniques or technical practices, usually of a discursive nature, designed to persuade an individual or motivate them toward a particular action. In this definition, rhetoric is largely a pragmatic or even mercenary endeavor. In this sense, rhetorical practices operate largely outside of or after epistemological concerns; even if one is attempting to convince an individual or audience of a particular fact or belief, rhetoric itself is composed of a series of adaptable techniques of expression without epistemological "content" or that are deployed after a particular fact, belief, or action has already been decided on. In his mid-1960s seminar on rhetoric, Roland Barthes gives a concise description of this conception of rhetoric in defining it as a "technique": "the art of persuasion, a body of rules and recipes whose implementation makes it possible to convince the hearer of the discourse (and later the reader of the work), even if what he is to be convinced of is 'false'" (13). This variability and adaptability, rhetoric's identity as a series of persuasive techniques, provides what Steven Mailloux has called rhetoric's "contingent universality": while persuasion may be a universal process in communicative acts of any type, and rhetoric as a discipline has largely been devoted to crafting and refining generic techniques that can be adapted to almost any situation, the contingent nature of both rhetoric and the contexts in which it will be deployed are already built into the process. In other words, while the lion's share of particularly early work in formal rhetoric has been devoted to the identification of techniques that might work in more or less any situation (or any situation of a particular type), the necessity that these techniques be flexible and that a rhetor adapt to particular audiences and situations has equally influenced the design and formalization of such techniques. Rhetoric has a certain "metastability" rather than a more traditionally static systematicity: particular rhetorical techniques have universal application insofar as adaptability and flexibility is a constitutive part of their design.

The notion that it might be considered as separate from any particular epistemological content has long been the source of rhetoric's vast range and adaptability, and in many historical periods, also the factor that justified its place in educational curriculum as a basic and crucial skill. However, this same separation from categories of veridicity or epistemology—particularly the notion that rhetoric, as Barthes notes, seems equally at home in being used to persuade people of something "false"—has also long been the source for the condemnation of rhetoric or a general mistrust of individuals who appear to be particularly good at persuasion. On the one hand, it is easy to dismiss contemporary rhetoric in the public sphere as dedicated more toward the maintenance of the status quo or at least to find its most apparent uses in the defenses or justifications of already decided-upon principles or conclusions, rather than acts of true deliberation or decision making. J. Michael Sproule gave the name "managerial rhetoric" to persuasive acts of these types, ones that appear to function largely as a way to disseminate and further ingrain existing beliefs and cultural tendencies within respective groups. Perhaps even more problematically, the variability of rhetoric as a technique is largely the inspiration for its appearance as a popular pejorative term, one that describes "empty" communication with no real meaning or merely the secondary, and often duplicitous, way to express the "real message" hidden underneath it. Both of these conceptions have long histories, with the idea that rhetoric is a dangerous pretender to the throne of philosophy or true reason, or that is only a lamentably necessary skill for the dissemination of knowledge into popular contexts, appearing as popular themes in Plato's dialogues. Both, however, also remain recognizable in contemporary culture, particularly within populist political contexts. For instance, the *Washington Post*'s "Fact Checker" feature, which awards a number of "pinnochios" based on the degree of deception manifested in recent statements by politicians, carries the tag line that it shows "The Truth Behind the Rhetoric." During their televised debates in the 2008 U.S. presidential election, candidate John McCain continually warned viewers to beware of Barack Obama's "eloquence" and use of persuasion, suggesting that is was being used to cover up the real intent behind his campaign promises.

The "technique" conception of rhetoric, particularly during the so-called "linguistic turn" in the humanities and social sciences and the rise of poststructuralist philosophy and postmodern cultural theory, was contrasted with one in which rhetoric was instead positioned as a crucial vector in the shaping of epistemological frameworks, and persuasion, as well as acts of communication and representation more generally, were taken to be in many ways constitutive forces in the production of knowledge. The theoretical position that was sometimes referred to as "Big Rhetoric" held that, insofar as virtually all categories of socially knowledge or belief—from science, to politics, to morals—seemed to be essentially mediated by acts of language and persuasion, then rhetoric appeared to be both the force that creates and subtends knowledge, and thus a privileged domain for analyzing how knowledge is formed and disseminated.[5] Though this viewpoint was certainly informed by a variety of contemporary changes in culture, politics, and technology—ones that often went by the proper name postmodernism—this contention is equally apparent (or at least became newly recognizable to modern audiences) in early debates over epistemology and persuasion in ancient Greek culture.

My own intervention in reference to these positions is to argue in this book for something of a recuperation or retrofitting of the first, technique-based, conception of rhetoric, but one that allows it to maintain much of the importance of it in reference to contemporary ethics and politics that was prevalent during the linguistic turn in cultural theory. More specifically, one of the major arguments of this book is that the power and importance we used to attribute to language and representation in the past few decades is, after the integration of post-cybernetic technologies into contemporary political economy, large carried out by what would more properly be considered as "techniques." By techniques I mean to refer to flexibly responsive practices that are directed toward motivating the performance of a generic action and/or the maintenance of a general equilibrium. One of the primary arguments of this book is that such techniques have largely taken the place of processes we previously attributed to epistemology or ideology within social life. I take up this issue much more extensively in Chapters 1 and 2, but to anticipate an example in the former, we might consider just-in-time production methods and niche marketing as particularly exemplary instance of this shift. If our traditional conception of consumerism under contemporary capitalism is one in which the producers and sellers of particular goods must work toward convincing a sizable amount of the populace that they should own such goods, then new production and marketing methods developed around information technology and new media have fundamentally altered this process. The ability to produce small batches of goods both cheaply and rapidly, combined with historically unparalleled ways to directly query consumers or obtain massive amounts of data about consumer preferences, creates an economic situation in which the primary goal is not to change a person's mind about the "need" for a particular product or service, but to instead create and produce virtually anything that is already desired, and to convince consumers to buy *any thing* at all. Insofar as we tend to identify capitalist exchange with forces of social normativity, such a shift also has much broader cultural effects. As Hardt and Negri suggest in *Empire*, far from being the enforcer of some engine of conformist identity or behavior, contemporary capitalism thrives by discovering that "every difference is an opportunity": "ever more hybrid and differentiated populations present a proliferating number of 'target markets' that can each be addressed by specific marketing strategies—one for gay Latino males between the ages of eighteen and twenty-two, another for Chinese-American teenage girls, and so forth" (152). While not entirely reducible to new technologies and media, such shifts, I argue here, are of a piece with a broader series of cultural changes occurring in the wake of cybernetics, ones in which techniques and hybrid categories of "technologics" increasingly play a crucial structuring role in social life.

There are a number of reasons that rhetoric provides the best perspective and best toolbox for analyzing and responding to these shifts beyond its historical linkage with *techne* and a particular logic of techniques.[6] While making those linkages is an objective consistently addressed and performed throughout this book, here we might briefly emphasize two points. One is that rhetoric has long been taken to be a vector of forces or practices that are, much like a variety of central processes in contemporary culture, premised somewhere between the application of physical force and the immaterial realm of pure reason or judgment. As the philosopher Michael Naas has emphasized in tracing its early thematization in ancient Greek intellectual culture, persuasion first emerged as "a mysterious third term that is properly neither overt, physical force nor reasoned, lawful judgment and compromise" (2). While, as alluded to above and suggested by Naas' definition by negation, it has been historically difficult to conceive of a hybrid space or "third term" between these realms, it is, I take it, this domain that structures a wide number of important processes in contemporary politics and culture, and it is the work of this book to thematize it using rhetorical theory. To put things perhaps a little more simply, we might say that rhetoric is particularly salient domain for analyzing contemporary culture because it, like the dominant processes of culture today, is less concerned with representation, epistemology, or ideology than it is with a spectrum of directly motivational or persuasive forces. As John Muckelbauer suggests, "because of rhetoric's traditional concern with persuasion (rather than communication), it has been intimately involved with questions of force rather than questions of signification or meaning," and that connection or involvement is perhaps more apparent and more important today than ever before (13).

In taking up ethics as my other primary toolbox here, I am particularly interested in its thematic and methodological overlaps with rhetoric that were most prominent in early Western intellectual culture: the ways that both are subtended by, or otherwise find their most productive strategies through the use of, common customs, investments, and desires. In ethics, emphasizing the "common" as both ordinary and the communal finds its roots in the Greek concept of *ethos* ("accustomed place" and the "character" of an individual), which preceded more formalized studies of ethics and morality. Thus, as Michael Halloran reminds us, "in contrast to modern notions of the person or self, *ethos* emphasizes the conventional rather than the idiosyncratic, the public rather than the private" and having effective ethos requires the ability to "manifest the virtues most valued by the culture to and for which one speaks" (60).

This primordial mingling of ethos as both communal custom and individual character persists in the two most prevalent ways that ethics is taken up in the contemporary humanities and social sciences: as (1) the study of human subjectivity or identity and how standards of valuation and judgment are formed and (2) investigation into the (proper) relations or activities between individuals within a given relationship or community. However, while the return of ethics in recent critical thought often begins with a consideration of the commonplace customs of a given community, positive ethics are more typically defined against such practices: ethical subjectivity and ameliorative ethical praxis are suggested to emerge, more often than not, through their difference or resistance to dominant communal sentiments and ideational investments. Rhetoric's genetic investment in the "common," by contrast, can be traced to the study and strategic use of *doxa* (forms of common belief and opinion). The sophistic commitment to *doxa* would be the primary source for Plato's condemnation of the Greek sophists and in many ways the linchpin through which rhetoric and philosophy would become, and remain, divided in Western intellectual thought. Thus, though these disciplines may out of necessity have to be central to any study of cultural and political processes, my own sense of rhetoric and ethics is tied to the ways their function emerges from the forms they take in relation to the "common": the strategic manipulation of the conventions governing both persuasion and our conceptions of justice, moral value, and "the good life." More specifically, insofar as contemporary forms of social power are increasingly premised on the manipulation of forms of persuasion, manipulation, and intervention rather than the particular content of these operations, it seems to me that it may be necessary to recover "formulaic" or generic senses of rhetoric and ethics in order to do any kind of strategic thinking about these forces. Taken together, such a conception of ethics and rhetoric, as well as their intersection, suggests that our best tools for responding to the present will be those that work via such persuasion to create unexpected linkages between individuals, particularly those who might not otherwise have common cause. Or at least it is of this that I hope to persuade you.

1 The Transhuman Condition

What I propose in the following is a reconsideration of the human condition from the vantage point of our newest experiences and our most recent fears.

—Hannah Arendt, *The Human Condition* (5)

The same month I began research for this book (November 2004), the UK's Human Fertilisation and Embryology Authority granted individuals with a family history of cancer the right to strategically select embryos free of these genes for fertilization.[1] As expected, this action promptly reinvigorated public debates over genetic engineering and the possible mass production of "designer babies"; critics of the decision have claimed that such selection not only is an unethical encroachment of technoscientific practice on the "natural" process of human reproduction, but that it also anticipates a future where unaltered humans are outpaced by their genetically enhanced counterparts and where "gene races" develop between countries (and corporations) to create and control technologies for producing "better" humans.[2] During the same month in the U.S., members of the National Academy of Sciences were working against a February deadline for their recommendation report on the legal and ethical status of transspecies chimeras, hybrid creatures created by implanting animal (including human) stem cells at the fetal stage into the member of a different species.[3] The Academy's report, a series of advisory guidelines for the federal government, will make suggestions regarding the regulation of contemporary chimeras—such as mice with human brain cells or pigs bioengineered to produce human blood—as well as investigate speculative matters, such as which civil rights should be granted to a hypothetical (but genetically plausible) human/chimpanzee hybrid (or "humanzee").[4]

Meanwhile, in Osceola, Wisconsin, legislation and contemporary technoscience were converging on a much more personal level as Elizabeth Woolley weighed options regarding a lawsuit recently threatened against the multinational Sony corporation.[5] Woolley's son Shawn, who had been previously diagnosed with clinical depression and schizoid personality disorder, committed suicide on Thanksgiving Day 2001. What made Woolley's demise a topic of popular media attention as well as a potential civil suit was the combination of his psychiatric diagnosis and his primary pastime: twelve-hour stints participating in Sony's online role-playing game

EverQuest. While the distribution of agency and culpability that would likely anchor a lawsuit is unsurprising—*your* video game killed *my* son—subsequent considerations of the matter provoked more complex questions. Insofar as such prolonged immersion in *EverQuest* would alter Woolley's dopamine levels, was his "obsession" a form of self-medication that eventually failed? Or did Woolley's intense engagement with a simulated reality hasten the acceleration of his neurochemical condition? In concert with coterminous research into the physiological effects of virtual realities and video gaming, speculations over Woolley's death left legislators, lawyers, and industry personnel scrambling to rethink legal and ethical guidelines to account for both the subjective and neurological impacts of electronically mediated experiences.[6]

Indeed, also in November of the same year, the potential affinities between narcotics and electronically mediated experiences were being considered in a congressional session sponsored by Kansas senator Sam Brownback.[7] Senators taking part in the session listened to a variety of researchers, such as Mary Anne Layden, co-director of a sexual trauma program at the University of Pennsylvania, argue that the effects of prolonged exposure to Internet pornography could replicate both the addictive properties and physiological effects of opiates such as heroin. Although critiques of pornography along moral lines or based on its supposed connection with criminal behavior have a long history, the combination of a relatively new delivery system (the Internet) and contemporary research into the neurological affects of electronic media on the brain are reshaping these debates. In this novel arrangement, anti-pornography advocates critique the consumption of pornography not (or at least not only) in reference to its presumed negative social effects or its potential to "corrupt" the morality of the viewer, but also for its ability to impact the brain in a directly physiological fashion. Hence, the question of controlling or censoring pornography becomes not so much a confrontation between moral values and freedom or speech, but an issue of public health or a corollary sortie of the War on Drugs.

Finally, November 2004 was also the month of the 55th U.S. Presidential Election and concomitantly the occasion of another novel intersection of politics and information technology. This particular election is probably best currently known as having occurred between *two other* elections of historical importance—the bitterly contested 2000 election that prompted the first intervention of the U.S. Supreme Court into the electoral process, and the election of the first African American president in 2008. However, it is quite likely that it will take on its own historical importance as marking the first time that sophisticated data-mining and niche-media techniques played a significant role in political campaigning. While the use of Internet and social media technologies for fundraising and volunteer organization was a widely covered phenomenon in this election, the more novel and pivotal integration of computing technologies was taking place largely behind such scenes. Specifically, 2004 was the first election that witnessed the significant influence of a number of "microtargeting" and data-aggregating firms that leverage massive server power and the increased availability of demographic information to identify for campaigns incredibly minute categories of potential voters and the best messaging strategies for winning their support. As journalist Steven Levy reports, the "fuzzy cohorts" developed around such categories as "family values voters" or "pro-choice voters" gave way in 2004 to the much more specific categories based on documented ideological and affective investments: a cache of "education obsessed Hispanic" moms microtargeted by Republicans and one of "Christian Conservative Environmentalists" wooed by the Democratic National Committee.[8] While there has always been a certain element of feedback between "polling and platform"—between what information a candidate obtains about voters' desires and what principles or causes they claim to be standing for—in political campaigns, the introduction of such sophisticated and precise methods for aligning these vectors seems to mark a more fundamental shift: mimicking in many ways the methods through which "just-in-time" or flexibly specialized production takes place in industry, it now seems possible for politicians to be dynamically and multiply shaped and presented in rapid response to the desires and investments of the populace they are addressing.

In this chapter I argue that these phenomena are representative of two broader shifts apparent in contemporary culture, both of which might be best initially grasped as extensions or intensifications of earlier and more recognizable cultural trends. The first is the way in which one of the constitutive qualities often attributed to the experience of modernity itself, the generally quickening pace of social change itself—what leads Paul Gilroy to refer to modernity as "the changing same" and Marshall Berman as "a struggle to make ourselves at home in a constantly changing world" (6)—extends beyond social or experiential factors into more explicitly material and biological realms. In a certain sense, the idea of modernity is by definition something like a permanent condition; as numerous critiques of early theorizations of *postmodernism* were quick to point out, what made the idea of modernity "modern" was the very notion that societies were entering a kind of permanent transition or rearrangement of cultural commonplaces and traditions in response to processes of urbanization, industrialization, and globalization. While this general case may remain true, we might suggest that the stakes and mechanisms of transition and modularity have changed over the past few decades; the modernity of "today" is different insofar as it would appear that nature rather than culture, or technology rather than ideology, are the vectors most visibly open for alteration and that play decisive roles in contemporary political economy. While it has been suggested that the present is an era of *hypermodernity* insofar as it seems like a "faster" or more intense version of our earlier understanding of modernity, perhaps it would be better to think of it as a time of *hypomodernity*, with the prefix "hypo" indicating a movement underneath or beyond the typical boundaries of what is in flux or open to alteration. As seen in many of the instances above, in such a transition questions of ethics and social policy typically asked in reference to human identity and the use of human reason become repeated on another level on the basis of human biology and our embodied capacities.

Second, we see a corresponding change in the targets and resources of institutions of social power. If, as Michel Foucault taught us, what distinguished the peculiarly modern form of social power that he called discipline was the ways that human "potentiality"—the various possible identities and physical behaviors of an individual—were integrated into its functioning, we might say that contemporary social power is focused on a broader category of human and non-human vectors that we might refer to as *capacity* or *capacitation*.[9] By this term I mean to emphasize the priority of not so much particular identities, or even what we might traditionally think of as categories of behaviors, but more general actions and tendencies of any type, such as the way in which consumer desires or political preferences of any stripe become grist for the mill of niche-marketing and niche-messaging strategies.

The following takes up these two tendencies as the appear within four features of contemporary culture that I take to be constitutive of what I am proposing to call "the transhuman condition" of the present, four different ways in which communicative or representational domains seem to have crucially intersected or become inseparable from more recognizably technological or physically material processes: a relatively recent change in the guiding objective of scientific research and production, one in which the epistemological pursuit of better understanding the natural world seems to have been overtaken by an emphasis on directly simulating or intervening in natural processes; new forms of capitalist production and marketing in which it is increasingly more difficult to definitively imagine any human activity as taking place outside of the realm of commodification; the ways in which human affect, the precognitive domains of human disposition or "feeling," have taken on new importance within political economy in conjunction with contemporary information technology and communicative media; and, finally, how a variety of related changes in contemporary politics and culture seems to have made the question of "humanism," and its limitations as an ethical framework, newly relevant in the present.

I will additionally suggest that each of these four phenomena tend toward a specific anitimony, in Immanuel Kant's canonical use of the term to describe the contradictory conclusions that seem to be mutually apparent within a particular state of affairs. However, here too we might already be one step removed from Kant's situation; as we shall see, Kant's partial solution to this problem—a consistent skepticism toward transcendental schemes of representation or valuation—is very much already itself a commonplace within the contradictions of contemporary cultural life that will be under review here.[10] As a continuation of the discussion of this antinomy of contemporary humanist thought in particular, I will end by suggesting how the kind of work usually performed via cultural theory that has been reliant on "representation" as a core category might be rethought for the demands of the present.

THE END(S) OF SCIENCE

The penultimate decade of the twenty-first century saw the publication of two volumes jointly titled "The End of Science." Though separated by only five years, these two texts contain very different rationales for making such a bold claim during a time in which science, or at least the industrial and technological products of scientific research, seemed to be having an increasingly larger impact on the daily lives of individuals in virtually every part of the world. Noted science journalist John Horgan's *The End of Science: Facing the Limits of Knowledge in the Twilight of the Scientific Age* forwards the argument that science has become, in a sense, a victim of its own success. Updating a 1960s prediction by molecular biologist Gunther Stent that we were on the precipice of a "golden age" of scientific discovery that would paradoxically result in an "end of progress" or rapid decrease in unexplored territory, Horgan suggest that this future has begun in the present across an increasingly larger share of scientific disciplines and fields of research. While the importance of existing scientific knowledge, to say nothing of its ubiquitous applications in contemporary society, remain prominent, the role of science as, in Horgan's words, "the primordial quest to understand the universe and our place in it" has ended, having become at its best an objective that promises "no more great revelations or revolutions, but only incremental, diminishing returns" (6). The other text under review here carried the less certain (and much more agonistic title) *The End of Science? Attack and Defense.*[11] Composed largely of the proceedings of a conference on its titular question held in 1989, the volume collects essays by both humanist academics and research scientists on what organizer Richard Q. Elvee describes as the ongoing re-examination of scientific disciplines "as a product of such things as paradigmatic focuses, ideological struggles and the basic instruments of power" (x). In place of the epistemological limit proposed by Horgan, here we have a political one: a greater understanding of the large variety of injustices promulgated under the guise of objective reason, the growing admixture of science's pursuit of knowledge with the goals of governments and corporations, and the emergence of "science" as a topic of research and critique by humanists, have led to both a new reflexivity within scientific representation and a growing suspicion of its post-Enlightenment role as a master discourse of legitimation.

The two different "ends" presented in these work are certainly distinctive; one might even think of them as oppositional. Horgan's claim for an "end" to science only makes sense if its previous discoveries remain sacro-sanct, whereas the closure suggested by at least one side of the "attack and defense" named in *The End of Science?* would appear to suggest the oppo-site conclusion: because our faith in scientific objectivity has been shaken, its discoveries are at least reopened for amendment by a perhaps chastened, but definitely more reflective and politically sensitive, array of scientific dis-ciplines. Nevertheless, here I want to pursue the possibility that at least the broad strokes of both of these claims about the near-future state of science have been proven prescient, and that our best opportunity for understand-ing the most significant changes in scientific enterprise over the last half-century or so resides in combining them in some way.

In doing so, however, I also want to complicate things a bit further by reading the "end" included in both titles as suggesting not only a conclu-sion or stoppage, but also a fundamental shift in objectives, a change in the "ends" or goals that seem to be most prevalent in scientific research and praxis (and, as we shall see, perhaps of contemporary "objective knowl-edge" itself on a larger scale). We might find a starting point for the first of these dual considerations in Horgan's qualification that applied science research has certainly not seemed to be under the same limitations as the more "primordial" quest for knowledge behind research in fundamental science (or what is often called "basic research" in the sciences). Indeed, on another reading, one might suggest that it is not so much that all of the major enigmas relevant to the vocation have been already "solved" by fundamental science at it is that there has been a shift in what "counts" as enigmatic, or what questions or challenges are worth pursuing within that domain. In other words, the case may be not that "fundamental science" reaches an "end" by answering all of its important questions, but rather that these questions have themselves become largely unimportant to the scientific, governmental, and corporate entities that increasingly prioritize direct intervention into the control of, rather than more accurate descrip-tions of, natural phenomenon and physical forces.

One could rather quickly account for—and if so inclined lay blame for—such a shift in the changes in funding priorities by governmental agencies following the heyday of so-called "Big Science" during and shortly after World War II and the allocation of private monies for the same pursuits. A 2008 study by the policy-setting body of the U.S. National Science Foun-dation, the National Science Board, approximates that of the around $340 billion spent on research in 2006, only about 18% was dedicated to basic research ($62B), compared with 60% to development ($204B) and 22% ($75B) to applied research. The difference is even starker if one considers only research and development funded by private industry, wherein money devoted to basic research has largely hovered below 4% of total research expenditures since the late 1990s. In early 2012, international debates over the outpacing of public funding by private corporations erupted following proposals made by UK science minister David Willet for a new class of science-centered universities funded entirely by the private sector, and by Minister of State for Science & Technology Gary Goodyear for Canada's National Research Council to shift its traditional focus on supporting basic research to what he referred to variously as "the business end," "the applied end," and "the commercialization-successful end" of scientific work.[12] What is perhaps most interesting about such controversies, however, has been subsequent arguments made *against* such proposals and the general decline in funding for basic research; proponents for increased public funding of basic research were all too quick to remind us of how "blue sky" scientific investigations undertaken without specific applications in mind have quite often turned out to produce significant new technologies and novel medical and industrial procedures. In this sense, then, even recent defenses of fundamental science or basic research seem to highlight its value as the precursor to its ostensible competitor, positioning fundamental science as valuable only insofar as it aids or leads to applied science and the development of new technologies and techniques that are immediately valuable to business and industry.

What is most striking here is not that scientific work has become increasingly commodified or beholden to for-profit entities—a historical sequence that one could of course quite easily trace throughout a wide variety of domains during the same time. Rather, what is peculiar is the way in which this shift demonstrates a certain inversion of the historical priority of scientific epistemology (over its technical applications) that had largely guided the public face of science, and provided it most visible social role, for several centuries of Western culture. In other words, what might unite the two "ends" of sciences proposed in the texts with which we began is that while one emphasizes the ideological role of science as an enforcer of certain standards of normativity and the boundaries of what counts as objective knowledge, the other bears witness to the disappearance of this role and the prioritizing of direct application and intervention as the core goals of scientific research and knowledge production.

Indeed, we might posit an even broader collapse in the boundaries between science's role as describing versus altering physical environments and elemental forces. Physicist and Science Studies scholar Evelyn Fox Keller notes, for instance, that contemporary molecular biology is becoming a field in which "the distinctions between representing and intervening, and more generally, between basic and applied science, are daily becoming blurred" through a kind of "conceptual instrumentalism" in which symbolic models of entities and processes are designed less for their representational accuracy or explanatory function and more for their pragmatic use in laboratory procedures and other concrete applications (S73). The pragmatic emphasis of such a shift, however, has by no means diminished the cultural, political, or ethical import of scientific work. As Keller, a veteran of the humanist critique of scientific epistemologies and practices, argues, it makes such questions perhaps even more urgent; the difference is perhaps that the blunt purposes behind desired "applications" for scientific research have now taken center stage as objects in need of interrogation, as opposed to the setting of normative standards for objectivity that were historically the focus of scientific discovery and discourse.

All of which brings us to the second dual consideration of "ends" introduced above; given such changes in scientific practices and priorities, what can we make of the other "end" of science proposed near the close of the twentieth century, the presumption that increased suspicion or criticism of science's claims to objectivity might terminate its social-epistemological function, or force it to assume another purpose in the cultural realm? Before too hastily concluding that all of these "ends" have ended up being the same—that the shift in scientific priorities from a normative or ideological function to a wholly pragmatic one may itself signal the end of science as culturally conceived through the majority of its history—we might do well to hesitate a moment over the recent history of the critical discourse that inaugurated attention to this second "end" of science, the sustained critique of science's ideological role within broader systems of social power. In other words, insofar as scientific epistemology became positioned as a synecdoche for a variety of ostensibly reductionist discourses and cultural commonplaces that came under attack by left-oriented academics and critics in the late twentieth century—and scientific objectivity served as both a counterpoint and a privileged target for the new critical tools of the social sciences and humanities—it seems worth asking after how the "withdrawal" of science from this role might be taken to reflect broader changes in social power or systems of cultural motivation and persuasion today.

Recall that the new politically charged and inherently skeptical attitude of such movements largely got off of the ground via distinctions between their own conceptual frameworks and that of the ostensibly disinterested theorizing performed in the "hard" sciences. Indeed, this is the path taken quite explicitly by what is undoubtedly one the founding documents of the enterprise, Max Horkheimer's manifesto-like essay "Traditional and Critical Theory." Beginning with what functions as most people's default conception of "theory," that of the natural sciences, Horheimer proceeds to show how the presumed ahistorical or objective nature of scientific theory leads it to collapse into a "reified ideological category" all its own, and then promotes a new kind of theory—"critical theory"—that might be a robust mechanism for mapping or revealing the very process through which history and social power is hidden behind the veil of transhistorical objectivity (194). In many ways this initial contrast—the definition of critical theory against the "traditional theory" of the sciences—would remain a constitutive feature of the enterprise; from "next generation" Frankfurt school theorist Jürgen Habermas' insistence that questions of the "good life" should remain the province of the humanities and of qualitative social science, to Foucault's critical historicization of the foundations or contemporary knowledge and power starting in sixteenth-century scientific discourses, the critical interpretation of scientific representation and its role in setting standards of objectivity has long been both an "acid test" for considering the range and power of critical theory's methods as well as a consistent target for its more explicitly political or pragmatic undertakings.

And it is in reference to this context that we might propose a distinct identity to the "political end" of science as considered in our source texts here, one that is certainly related to the triumph of praxis as science's primary driver, but perhaps poses its own dilemma about the social role of science today as well as the present purpose or "ends" of (critical) theory in the humanities and social sciences. More precisely, it seems that the displacement of science as the master discourse of veridicity or objectivity has been largely, and one would presume not purely coincidentally, coeval with a decline in the importance of "objectivity" itself as a central vector of contemporary political power. Indeed, it would even appear these days that the kind of negative hermeneutics pioneered by critical theorists around the mid-twentieth century—the very ones developed in part around the critique of the decontextualized objectivity traditionally identified with science—has in many ways taken on this role.

We can find a particularly appropriate example of the latter within a recent essay by Bruno Latour, one that reads in many ways like a pseudo-*apologia* for the discipline of Science Studies that Latour himself was largely influential in birthing and popularizing. Early in "Why Has Critique Run Out of Steam?" Latour provides a partial answer to his titular question by noting the recent appropriation of one of the discipline's major "moves"—the questioning of the objectivity of a scientific discourse in regard to the biases or ideological motivations of its proponents—for purposes far removed from those of the Latour and his colleagues. Detailing how (in)famous conservative pollster and political strategist Frank Luntz has challenged scientific studies of global warming as both uncertain and politically motivated, Latour ponders how such a scenario might have emerged from, and where it might leave, the left-oriented critique of science or epistemology in the humanities and social sciences over the past several decades. One the one hand, Latour notes, such a situation seems like a reversal of the one that provided the exigence of the critique of science many decades ago; the "the danger" at issue for the contemporary critical humanities and social sciences is "no longer coming from an excessive confidence in ideological arguments posturing as matters of fact—as we have learned to combat so efficiently in the past—but from an excessive distrust of good matters of fact disguised as bad ideological biases" (227). Given such an ironic reversal, and the very real risk posed by the appropriation of this kind of critique by retrogressive parties, Latour goes on to consider whether it might require something like an entire inversion of the critical enterprise as a whole: "while we spent years trying to detect the real prejudices hidden behind the appearance of objective statements, do we now have to reveal the real objective and incontrovertible fact hidden behind the illusion of such prejudices?"

Latour's ostensible *mea culpa* here is, of course, one of many entrants in a larger discourse around the "death of theory" or the need for a radical rethinking of the commonplaces and strategies of humanities and social science theory that have become prominent over the past decade or so. Latour's response is notable, however, for emphasizing both how larger concerns about the "death of theory" relate back to its origins in the critique of scientific objectivity, as well as for how it emphasizes that its ostensible decline—much like Horgan's reading of the decline of scientific discovery—is largely the result of its own success. Having been so persuasive for so long in its presentation of the ideological underside of scientific theory and the fragility of claims for pure objectivity of perspective, the critical left of the humanities and social sciences have found their tools now being taken and repurposed by the very "enemies" they were once designed to combat.

Taken together, then, these two "end(s)" of the traditional roles of science leave us with two distinct antinomies, both of which are bound up with important political and ethical questions. On the one hand, we may have to admit that the late twentieth-century critique of scientific objectivity—without retroactively depriving it of any of its achievements in foregrounding past and contemporary tragedies carried out in the name of science or alibied by the ostensibly disinterested search for objective truth—may have achieved its greatest success only when (or only because?) science was itself retreating from its role as the dominant provider of epistemology or normativity at the same time. If we, rightfully, I would think, consider that undertaking as only one of the most intense areas of a larger progressive agenda in critical thought and the analysis of modern social power, then we are left with a perhaps more thornier and more urgent question as well: if the broader challenging of normativity and the decontextualized commonplaces of dominant knowledges and practices—the action that more or less put the "critical" in "critical theory" from the mid-century onward—has now become equally at home in the rhetorical strategies of the most regressive and moribund movements of the present moment, then what now can play the role of "theory" today, what framework of analysis or contestation can act as a check on the more destructive or exclusionary systems of knowledge and power?

THE TECHNOLOGIC OF CONTEMPORARY CAPITALISM

As alluded to above, changes in the teleological anchors of contemporary science research, to say nothing of such controversies such as those over possibilities for genetically altering human bodies addressed earlier in this chapter, can by no means be separated from a larger shift in contemporary capitalist economics, one in which scientific research and such technoscientific process as the creation of "designer babies" or radical life extension might be viewed as only instances of a wider variety of material phenomena that were previously considered unassimilable to processes of commodification but have recently been made available as a product or service to be "consumed" in contemporary economic exchange. While the actual commodification of aspects or states of the material human body might be a fairly radical example, much recent work distinguishing contemporary capitalism from its earlier forms has emphasized the less extreme incorporation of a number of human cognitive or affective capacities into the realms of capitalist exchange. More specifically, if one of the key features of global capitalism of the 1980s was its emphasis on privatization and the expansion of commodity categories—notable, on the one hand, in the transfer of publicly held goods and services to private corporations and, on the other, in the general rise of a variety of "immaterial" goods and services for sale—the capitalism of the 1990s onward seems like something of a extension of this trend in the opposite direction—one in which the "labor" or "production" side, rather than that of consumption or commodities, seems to have undergone a similar transformation.

For these reasons analyses of these changes have tended to focus on the ways that what we might count as "labor" in the most general sense—human activity that produces recognizable "value" within a particular economy—has fundamentally changed shape during the past several decades. Recall, for instance, that two of the greatest buzzwords of '90s management theory and economic forecasting were "knowledge work" and "symbolic-analytic work"; both terms attempt to encapsulate the value produced by activities that are largely cognitive, or based on competencies in analysis, problem solving, and strategic communication—tasks that seemed to be increasingly at the center of the so-called "New Economy" of the late twentieth century. Around the same time, a similar focus became dominant in much economic scholarship with a strong (post-)Marxist influence, but in this case expanded to include the ways in which "knowledge work" has increasingly become an additional responsibility of even manual laborers, as well as to chart how the "passive" contributions of knowledge or information by individuals not recognized (nor compensated) as workers are integrated into contemporary production processes. Here, a wide variety of terms have been proposed to name this new category of labor or the economic mode based on the intangible or passive "work" of individuals: affective labor, (Hardt and Negri), immaterial labor (Lazzarato), virtuous labor (Virno), linguistic labor (Marrazi), semiocapitalism (Berardi), spectacular-work (Beller), and cognitive capitalism (Moulier-Boutang). Whereas the more traditional analysis of market transformation named by the "knowledge worker" typically responds to the pragmatic questions of human resources management—how does one motivate and cultivate immaterial labor as opposed to the more apparent and quantifiable actions of workers? How do governments and other institutions better prepare their economies and their people to perform in these conditions?—these latter categories tend to update and extend Marx's labor theory of value, positioning cognitive work or immaterial labor as a new and frequently unacknowledged level of worker exploitation under capitalism. Thus, for instance, Hardt and Negri write of the increasing dependency of capitalist production on "a collective, social intelligence created by accumulated knowledges, techniques, and know-how," but not one that has necessarily created a more egalitarian or collective share of wealth or power than in earlier stages of capitalism (*Empire* 364). Indeed, instead it seems to have led to an even greater level of estrangement from one's own labor, and an even further siphoning of such "value" away from its original producer.

There is, of course, a particular tension, or implicit contradiction, in analyses of these changes that take their cue from Marx's analyses of labor in tracking the novel ways that *labor* has become an increasingly diffuse referent in the contemporary production of economic value. On the one hand, and as autonomist-influenced economic theorists in particular have been quick to point out, there is much in Marx that seems to anticipate or proleptically account for how modern labor has always involved a certain kind of recurrent exchange between the more subjective or affective capacities of the laborer and their actual "material" work. As Marx tells us early in the section of *Capital* specifically devoted to describing the "labor process," through it a worker

> confronts the materials of nature as a force of nature. He sets in motion the natural forces which belong to his own body, his arms, legs, head and hands, in order to appropriate the materials of nature in a form adapted to his own needs. Through this movement he acts upon external nature and changes it, and in this way he simultaneously changes his own nature. (283)

Similarly, while Marx's default examples of "laborers" largely remain field and factory workers, he quite consistently considers the rather different modes of production common to teachers, artists, and other prototypical "knowledge workers" of his own era.[13]

On the other hand, however, the fact that the immaterial or affective components of value production seem to be increasingly the rule, rather than the exception, within capitalist economies would seem to cause substantial problems for Marx's analysis of political economy in general, or of "value" as a whole. In other words, while Marx's interest in intellectual labor or in the nineteenth-century version of what we would today call the "service economy" is certainly prescient, the fundamental distinction between labor and other areas of social life and subjective experience largely remains the lynchpin of Marx's approach to everything from his theories of human consciousness, to his analysis of the internal contradictions of capitalism, to his suggestions for how political revolution can be planned and promoted. In particular, the ability to quantify labor, to evaluate how the surplus of value created by a laborer, that vector of her production that exceeds her renumeration and thus, as Marx writes, "for the capitalist, has all the charms of something created out of nothing" would seem to provide both the ethical and analytical core of his work: it is the fundamental distinction through which one can index the inequality of the capitalist system and anchor the entire hermeneutical sequence that would become known as historical materialism (*Capital* 325). Thus, while the centrality of labor in Marx's thinking would seem to make it a particularly appropriate framework for responding to a time in which "labor production" seems to bleed into ever more areas of everyday life, to do so would be to ignore how the *value of* that category is sustained by our ability to conceive it as recognizably separate from a variety of other domains of practice— consumption, leisure, the subjective experience of life "itself"—to which it might be opposed. All of which is to say, while extending our concept of labor to include cognitive and affective phenomena seems to concisely describe the various ways that labor has been disconnected from particular times and spaces as well as from immediately productive activities—from the middle-manager answering work e-mails from their home in the middle of the night, to the food services worker told, as I often was, that smiling is "part of the job"—it doesn't seem to adequately describe the full force of the changes subtending such processes, the broader distinctions between labor and almost every other category of activity that have been progressively undermined over the past few decades.

For these reasons, I want to pursue a somewhat different point of intervention here, one that similarly seeks to account for what is unique about the last several decades of socioeconomic exchange, but one that takes the broader admixture of epistemological or communicative capacities and of overtly material or technical forces, to be the primary, and perhaps even more vexing, focal point for these changes. In particular, it seems to me that the very same advances in, and increasing importance of, technologies and communicational media so apparent in the other areas under review in this essay are crucial considerations for any attempt to rethink the contemporary status of economics, let alone labor, in the present moment.

We might even locate the groundwork for such a technics-oriented analysis in the very same passage of Marx's *Grundrisse* that has inspired much current work on immaterial labor, the so-called "Fragment on Machines." Here Marx extends his analysis of how machinery used in production serves as a reproduction and replacement for human labor to reflect on the ways other human capacities are absorbed by capital:

> In machinery, objectified labour itself appears not only in the form of the production of the product employed as means of labour, but in the form of the force of production itself. The development of the means of labour into machinery is not an accidental moment of capital, but is rather the historical reshaping of the traditional, inherited means of labour into a

form adequate to capital. The accumulation of knowledge and of skill, of the general productive forces of the social brain, is thus absorbed into capital, as opposed to labour, and hence appears as an attribute of capital, and more specifically as *fixed capital*, in so far as it enters into the production process as a means of production proper. (694)

In autonomist-influenced economic criticism, this observation is frequently cited as a prescient description of the situation under contemporary capitalism, one in which labor is no longer constricted to particular sites such as the farm field or the factory floor and is instead distributed across a wide range of social activity. In this sense, the emergence of what Marx calls "general social knowledge" as a direct rather than ancillary force of production in capitalism (706), what is often called, after Marx, "the general intellect," is something like an extension of the ways in which physical machinery can be considered as "objectified labor" (in this case labor quite literally "made" into a material object, with machines acting as replacements for human labor in the production process as well as also being the material offspring of knowledge and intellectual labor). As Paolo Virno glosses this transition, if the integration of physical machinery into the common production processes of capitalism is one in which "knowledge is objectified in fixed capital, transfused into the automatic system of machinery and granted objective spatiotemporal reality," then the path traveled in the likewise integration of general social knowledge looks much like the same sequence without the mediating step that occurs when "knowledge" is embodied in physical machinery; instead communicational and representational schemas themselves, which Virno lists as including "theorems of formal logics, theories of information and systems, epistemological paradigms, certain segments of the metaphysical tradition, 'linguistic games,' and images of the world," have become the modern-day "machines" of the contemporary economy in a fashion that parallel Marx's analysis of the industrial machines of his own time, and in fulfillment of his predictions for the increased importance of the general intellect into emergent modes of economic production ("Ambivalence" 22).

However, here again it is important to separate what we might call Marx's functional prediction—that abstract and socially shared knowledge would become a primary productive force—from his explicitly political prophecy about that process: that this process would form what he calls a "moving contradiction" within capitalist economies, that the novel admixtures of productive forces and social relations would short-circuit traditional capitalist structures built around their separation (*Grundrisse* 706). Yet it would appear that between Marx's time and the present, very much the opposite has been proven to be the case. As Virno suggests, "what is striking now is the complete realization of the tendency described in the 'Fragment' without any emancipatory or even conflictual outcome" (23); the potential conflict between these two vector has instead become a baseline principle of contemporary capitalist production as a whole, the "moving contradiction" formed by the intermingling of sociality and economic production is now a major mover of the economy.

I want to suggest here that the root of this issue is not so much an extreme ambiguity between what falls inside and outside of labor as a recognizable category, but a larger intersection or overlap of these forces of material production and of representation or communication as a whole, in other words, the specifically economic appearance of the larger phenomenon under review in this text—the general conflation of technical and epistemological domains that has taken place over the last several decades. More specifically, we might say that the strange intersection or zones of indistinction between (explicitly) economic activity and social or generically human activity is largely the result of the ways in which contemporary technology and media have made the active or passive co-creation of commodities an intrinsic component of general economic production.[14]

From this perspective the real shift in contemporary capitalism as compared with a traditional (post-)Marxist analysis of political economy, would be that it is increasingly more focused on the purely technical operation of keeping economic circulation of any type functioning and increasingly less reliant on the need for "superstructural" enforcements to produce this effect (and thus less dependent on maintaining particular ideological or representational forms that might foment desires for particular commodities). At the same time, however, the symbolic or cognitive forces that used to be taken as forming such superstructural elements take on a strange new materiality. For instance, while Virno still tends to consider such newly apparent forces of production as extensions of traditional labor—the subsumption of previously "social" or "private" capacities of the individual into the realm of work—his description of them seems to already presuppose a more global rearrangement of communicative and representational capacities, ones in which they tend to take on an almost physical force. He writes, for example, of how the general intellect is an abstraction, but one "equipped with a material operability"; similarly, the various capacities of "general social knowledge" being put into play as forces of production are composed by "axiomatic rules whose validity does not depend on what they represent. Measuring and representing nothing, these techno-scientific codes and paradigms manifest themselves as constructive principles" (23). The larger point of contention or seeming contradiction within contemporary political economy seems to be not so much the collapse between labor time and what Marx called the "disposable time" in which individuals are not directly producing value within capitalism (*Grundrisse* 208), but rather the convergence of the technical and representational into a series of generic technics or techniques that dominate both the recognizable field of capitalist economic production as well as whatever we might have taken to be the "non-economic" component of quotidian human life (or as we shall see, even the "anti-capitalist" activity of individuals).

In addition to labor, then, we might also trace the appearance of such a convergence in the overlap or functional similarity between production and consumption, as well as the full-scale integration of a variety of modes of "non-economic" collaboration into capitalism. In particular reference to the first of these, we might return, by way of example, to the vast expansion of demographic research and marketing techniques alluded to in the beginning of this chapter. If our default conception of consumer capitalism is one in which individuals must be convinced to purchase available commodities irrelevant to their essential needs, it would seem that if changes in the ability to produce goods—particularly increases in flexible specialization and "just-in-time" production at the level of manufacturing—as well as to determine existing consumer desires and interests (through complex methods for collecting and indexing demographic data and consumer dispositions) have not changed the core principles of this system of exchange, it has at least substantially reversed the burdens placed on "creating" versus "responding to" the dispositions and desires of consumers of all stripes. To use a rather prosaic example, if the challenge of a U.S. refrigerator manufacturer in the 1960s was to convince consumers that they needed to purchase or upgrade to the new mass-produced model currently awaiting shipment from strategically placed warehouses, the challenge for the individual in the same position today is increasingly to determine the precise attributes of a refrigerator that would automatically appeal to a large enough (though increasingly smaller) market to justify producing a cost-effective amount of them. To account for such changes, one would have to rephrase Adorno and Horkheimer's paradigmatic description of consumer capitalism as the system in which "something is provided for all so that none may escape" (*Dialectic* 123); today, it would be more accurate to state "*everyone must provide something* so that none may escape": capitalist production and consumption does not work so much by providing a wide enough variety of options to project a fantasy of freedom of choice, but by the targeted response to almost any kind of desire near the moment of its very creation; and, for better or worse, declaring some kind of identity or desire is increasingly the cost of having any kind of participation in contemporary economics and politics, however quotidian (I have to allow my local grocery story to track my purchasing practices in order to take advantage of the "great deals" available through their membership card) or monumental (one must claim some kind of ideational or demographic position in order to be heard, or at least to "count," in the public political arena). Indeed, one might even go one step further and state that "*everyone must be something* so that none may escape": because everyone must have some kind, any kind, of subjectivity whatsoever, that subjectivity will be targeted and some economic value will be extracted from it.

Well beyond the realm of "traditional" manufacturing and marketing of this type, we might consider a much larger field of processes through which the distribution and organization of voluntary (and unremunerated) activity is incorporated into production processes on two other levels. First there is the relatively "active" participation that takes place in such processes as *crowdsourcing* (the outsourcing of portions of a larger task to an indefinite group of volunteers), *prosumption* (a consumer's direct participation in the production of the good or service being consumed), and *citizen journalism* (the coverage, documentation, and public discussion of new event), as well as in participatory and social media of all types. Second, we might consider the relatively more passive contribution of demographic data, geographic location, and consumer dispositions used to generate niche- and proximal advertising and messaging strategies, ones that are increasingly driving the co-crafting of, as we saw in the start of the chapter, both products and politicians over the last two decades. Taken together, these phenomena constitute a significant reordering of economic and communication networks, as well as systems of social power more generally. The emerging configurations in each draw their value and function through, on the one hand, the flexible *division* of materials, processes, and tasks that might be shaped altered, or fulfilled by leveraging the individual contributions of ad hoc communities, and, on the other hand, the broadest possible *inclusion* of individuals of all backgrounds into such collectivities.

As alluded to above, the importance of collectivity and collaboration within such practices is perhaps worthy of special consideration here. If one way to look at the expansion of value production is to track the ways in which individual capacities and behaviors previously taken to be constitutively *private*—in the sense that are not in typically part of the public realm of exchange—have become commodified or "monetized," then we might say this tendency has also been matched by a similar subsumption of collective and cooperative behaviors, in many cases precisely those that were taken to form the contrastive "outside" or edge of economics as the realm of the public exchange of value and the organization of cooperative activities. We can find a particularly salient description of this process in a recent work by business consultant and digital economy guru Don Tapscott (writing here with Anthony D. Williams). As Tapscott and Williams suggest, the "new forms of mass collaboration" such as crowdsourcing and peer production that "are changing how goods and services and invented, produced, marketed, and distributed on a global basis" are largely variations on the "old forms" of cooperation more common to ostensibly non-economic, and sometimes explicitly anticapitalist objectives and collectives:

> In the past, collaboration was mostly small scale. It was something that took place among relatives, friends, and associates in households, communities, and workplaces. In relatively rare instances, collaboration approached mass scale, but this was mainly in short bursts of political action. Think of the Vietnam-era war protests or, more recently, about the raucous anti-globalization rallies in Seattle, Turin, and Washington. Never before, however, have individuals had the power or opportunity

to link up loose networks of peers to produce goods and services in a very tangible and ongoing way. (10)

Though one might balk at the seemingly gleeful tone through which Tapscott and Williams seems to be celebrating new opportunities for businesses to exploit cooperative behaviors that used to be restricted to the "wastefully" unmonetized activities that united families, friends, and political activists, there is much to suggest that it is this analysis, rather than the one implicit in much post-Marxist scholarship on new economic formations, that most adequately presents the kind of full saturation of capitalist commodification within social life. Tapscott and Williams' reference to "the raucous anti-globalization rallies in Seattle" is particularly notable here, given how widely the so-called "Battle in Seattle," the protests of the World Trade Organization Minsterial Conference of 1999, has been held up as a particularly promising example of collectivity and mass protest against the excesses of contemporary global capitalism (Hardt and Negri, *Multitude* 285; Pignarre and Stengers 3–9).

All of which brings us to our third antinomy pervading contemporary cultural life: the increasing appropriation of "non-economic" activity into capitalism is not so much something like a "triumph" of that system, but a bellwether of the ways in which more generic techniques of managed collaboration and social interaction have been set free from any clear mooring inside or outside of capitalism as a formal system. Another way to put it would be that capitalism *isn't really capitalism* anymore, or at least no longer takes the form that we usually associate with the concept. This, I take it, is one point on which the more typically left-radical analyses of immaterial or cognitive labor and the more traditionally conservative analyses of "knowledge work" seem to agree, though they tend to map the disappearance of the distinction between capitalism and its others in opposing direction. For Virno, the post-Fordist realization of many of the core tendencies that Marx associated with the eventual decline of capitalism (and concomitant emergence of communism)—the rise of the general intellect as a productive force, the widespread availability of opportunities for social cooperation, the increase in global communicative structures—*without* the actual occurrence of that decline, leads to something best phrased as "the communism of capital," an integration of socialist practices and possibilities without the aggregative benefits that would presumably come along with them (*Grammar* 110–111). Moving in a different direction, the very same changes led to prominent management consultant and business guru Peter F. Drucker's declaration of an emergent worldwide "post-capitalist society"; for Drucker, writing in the early '90s, "the real, controlling resource and the absolutely decisive 'factor of production' is now neither capital nor land nor labor," but the more ethereal techniques associated with knowledge work and the post-industrial finance and "information economy" (*Post* 6). Going with it, Drucker goes on to suggest, is any clear consideration of capitalism as the "dominant social reality" of the time, its decline mirroring the end of socialism as a "dominant social ideology" in the Eastern Bloc states.

Viewed in this light, the emergent technicity of contemporary value production and social activity, its flickering state between the more formally solid domains of the purely representational and communicative and the more determinatively technical or "operational," is something like the metaphysical parallel to the more concrete and site-specific changes that have eroded the distinctions between capitalism and socialism during the past few decades: socialist countries' investment of sovereign wealth funds into capitalist economies and the related enlargement of labor union and pension fund investment in capital markets (what Drucker famously called "pension fund socialism" in an earlier work), the purchase of large amounts or full-scale takeover of publicly traded and private corporations by capitalist governments during and in the aftermath of the recent economic "crisis" of 2008 (*Unseen*). While these phenomena are fairly specific harbingers of the newly murkier boundaries between formal economic systems—one in which many capitalist economies has lost much of what Erik Olin Wright calls their "capitalisticness," while socialist economies have embraced many features of traditional capitalism (36)—they might be taken as only the more obvious signs of a growing parity between the techniques taken to be paramount to these forms, or more broadly as the emergence of "techiques" themselves, the largely neutral middle-state between explicitly representational and operational vectors, as the major force field through which contemporary culture finds shape and coherence.

All of which leads us with a number of pressing questions about how one seeks to either conceive of, let alone intervene in some oppositional way within, the economic in any way that would do something other than merely reproduce or further extend its already thick saturation of the social. Insofar as the economic has now come to encompass a variety of generic human capacities and quotidian social relations, then how can one attempt to withhold their value production from this expanded field? If one produces value of some sort every time they click on a link in a search engine or gaze upon an advertisement, then how does one opt out of such actions without removing themselves entirely from participating in any type of social exchange, or by refusing to exhibit any of an increasingly larger share of basic human capacities and behaviors? Indeed, for theorists who take the expansion of labor into immaterial realms, the question of "opposition" or "resistance" to such forces has often been met with what we might take to be equally absolutist strategies of negation or disassociation: considerations of how one might stage an absolute defection from current social conditions or rigid resistance to commonplace modes of behavior and interaction.

We might find a starting point for this tendency in Mario Tronti's influential early '60s essay "The Strategy of Refusal," which argues that "mass passivity at the level of production is the material fact from which we must begin" if we hope to develop anything like a viable resistance to emergent forms of capitalist exploitation (20). More recently, Virno has hypothesized resistance to the capillary tendency of labor around a series of references to more extreme or fantastical variations on the factory strike or "walk out," a search for a form of "exit" or "exodus" from the matrix of capitalist value production as a whole (*Grammar* 70–71). Finally, in acknowledging the increasingly biopolitical nature of contemporary multinational capitalism—its abilities to extract value from almost any aspect of human functioning—Hardt and Negri suggest that any "will to be against" the increasingly larger enemy of global "empire" would require us to inhabit "a body that is completely incapable of submitting to command . . . a body that is incapable of adapting to family life, to factory discipline, to the regulations of a traditional sex life, and so forth" (*Empire* 216). While the diagnoses behind such suggestions can hardly be accused of understating the significance of the extension of value production from defined categories of labor to overwhelmingly common elements of generic social life, their prescriptions for redress or resistance share an overwhelming negative or passive tone—suggestions of what one *should not do* in order to inhibit their exploitation within this system, rather than what one could or should do in an attempt to redirect or ameliorate it in some way.

We might then all too hastily conclude this swerve through the economic valences of the technologic of contemporary culture by suggesting a more refined variation of the antinomy offered above, a paradox or constitutive tension less diagnostic than it is strategic. Insofar as the novelty of contemporary economic production and circulation seems to be the cleavage of techniques from the ideological or ideational domains with which they were previously associated (the socialist or the capitalist, the public or private, the competitive or the collaborative), the fundamental challenge of the present is not so much to discover some radical alternative to contemporary conditions—one that often circulates rather nebulously in notions of an absolute refusal or exodus—but to figure out how these same techniques already immensely immanent in contemporary capitalism can be made to produce different outcomes, to somehow ameliorate the immense inequalities or material damages that largely remain common to the system, despite its vast mutations in other areas. In other words, the real lesson offered by the citation of anti-globalization protests as a "business model" for contemporary corporate entrepreneurs is that the prevailing technicity of contemporary culture has made a large number of techniques up for grabs by partisans of diverse and conflicting agendas and objectives. Under the "communism of capital" or in "post-capitalist society," the trick is to acknowledge and in some way leverage the fact that these techniques will always already be commodified or draw their value against the backdrop of mechanisms for modulating desire and motivation that we have traditionally come to identify with capitalism. This suggestion itself, of course, perhaps raises its own questions about both the pragmatics and ethics of such a strategy, one that might be at least provisionally approached around the consideration of affect, a force and concept that has itself often been invoked in relation to contemporary political economy, and one to which we now turn.

HOMO-MACHINIC PANIC

Indeed, we might also find a constellation point for all of the earlier discussions in this chapter—not only the changing role of economic value production, but also in relationships between technoscience and politics—in the increased attention to the central role of affective or sub-rational forces in contemporary culture, and what has been deemed the "affective turn" in contemporary cultural theory that has arisen in response to it. In both instances, the increased availability of technologies for measuring human physiological response has resulted in a fundamental reconsideration of the ways in which our embodied capacities and dispositions (perhaps even more so than our thought process at the level of consciousness) have become central to contemporary politics and culture. Here too we are faced with a situation in which ostensibly acultural or transhistorical properties (i.e., the human nervous system) appear to be taking on new forms or functions in response to novel changes in culture. Similarly, as I will argue below, this reformulation of the connections between the natural and cultural may present some particularly urgent questions about the intersections of rhetoric and ethics in the present moment.

The root causes for the turn to affect in contemporary cultural theory, or the more general feeling that it is affects rather than ideational content that seems to drive our contemporary experiences, might be mapped across a number of cultural domains. Perhaps most notably, renewed interest in affect in the humanities and social sciences has been largely coeval with the exploration of the role of affect in contemporary marketing research and practices. The widespread use of sophisticated diagnostic technologies for measuring individual's precognitive responses to advertising appeals, as well as the general affective associations consumers have for popular products, has been at the center of so-called "neuromarketing" over the last decade or so.[15] The use of functional magnetic resonance imaging (fRMI) and Electroencephalography (EEG) machines to measure brain activity in response to particular advertising images and appeals has, on the one hand, led to a greater awareness of the ways in which such appeals have always operated beneath the cognitive level and, on the other, made it much easier for those who design such messages to more strategically exploit these tendencies. If, as discussed earlier in this chapter, the "ideological" model of capitalist consumerism based on expectations of mass conformity has been replaced by niche and micromarketing that more flexibly targets small groups of consumers, then neuromarketing might be taken as a parallel movement—one in which even one's affective capacities are integrated into the production and selling of goods and services.

One might also, for instance, chart the rise of affect's perceived centrality to human experience against the increasing replication of other human capacities in mechanical realms; as we have witnessed a variety of capacities previously taken to be unique domains of humans produced within technological systems, affect has correspondingly been defined as property uniquely singular to humanity (or at least one restricted to certain set of "advanced" biological creatures). Similarly, one could index the turn to affect in relation to changes in technological resources and aesthetic styles of various media, such as cinema. In this genealogy, a line could be drawn tracing movement from the early "cinema of attractions" (e.g., *Arrival of a Train at La Ciotat*, 1896), to the birth of narrative film proper (*The Great Train Robbery*, 1903), to the era of pastiche and narrative recycling (to keep with the train theme: *Throw Momma from the Train*, 1987), and, finally, to a number of more recent films that stage a return of sorts to the cinema of attractions by targeting viewers' affective responses explicitly through nonlinear sequencing and the prioritizing of emotionally intense images (e.g., *Trainspotting*, 1987, clips of which have in fact been used in controlled experiments to prime particular affects in participants) (Lerner; Hall).

However, the shift that seems to be most pressing to the majority of researchers participating in the contemporary turn to affect might best be describe in reference to political economy, if we take the broad meaning of that term to encompass both how value is created within culture and the complicated mediations between individual and group identities within that process. This genealogy might begin with Adam Smith's paradigmatic gesture in theorizing "the invisible hand" of the market. It was Smith's invention that fascinated the young Hegel and inspired his conception of a "ruse of reason" to explain the complex interactions between the conscious and unconscious motivations of individuals and between the motivations of individuals and social collectives. This latter structure, and in particular its further transposition of Marx in his writing on ideology, is the one most commonly being worked through and against in contemporary cultural theory and criticism that suggest that our embodied and affective qualities significantly shape our political and ethical dispositions.[16] As a source for contemporary (cultural, political) motivations or investments, however, affective processes can seem all the more elusive for having a material "location" in the human physiology; indeed, such a location often makes affect appear both ubiquitous and comprehensively unaccountable, a replacement of the "invisible hand" of rational choice with the "invisible gland" of affective processes now driving political economy.

Indeed, it seems to be this last vector in particular—the distinction between the embodied and preconscious forces of affective response and our more traditional conceptions of cultural norms as socially constructed or ideologically enforced—that has been a major selling point for theories of affect today. For example, in his immensely influential writings on affect, Brian Massumi argues that increased attention to the role of human physiology can help us pass through the "gridlock" created by cultural theory's recent emphasis on identity and an individual's position inside an "ideological master structure" of social norms; for Massumi, shifting focus to the precognitive capacities associated with human emotion might allow us to avoid both the "Scyllan of naïve realism" and the "Charybdis of subjectivism," and thus capture more concretely the co-relation of human bodies and human culture (3–4). Similarly, Eve Kosofsky Sedgwick's equally prominent work on affect positions it as an opportunity to challenge what she calls the "antibiologism" of critical theory in the humanities and social sciences, and to think outside our tendencies, post-Foucault, to dedicate our scholarly efforts on categorizing social forces as either "repressive" or "liberatory" (101, 10). For both Massumi and Sedgwick, as for the majority of participants in the affective turn of contemporary cultural criticism, affective potentials are largely "hardwired" into human biology—the potentials for such affects as shame, joy, and so on, are fairly universal to human experience—so attention to its role in human behavior and identity is taken as a salutary way to rethink those categories outside of an emphasis on sociality as the prime shaper of human subjectivity and experience. In particular, focusing on the role of biological response in shaping human subjectivity and experience is forwarded as a way to complicate any notion that an "intellectual" understanding of sociality or the contingency of interpolative forces automatically provides some purchase on resisting them. In this sense, we might say that while the general interest in the role of affect is a response to changes in contemporary technology, communicational media, and aesthetic forms, the turn to affect theory is also an attempt to restage or reorient our more traditional methods of reflection and analysis: noting the increasing prevalence of "nonrational" modes of communication and persuasion helps curb our overly optimist, or one might say all too humanist, hopes that rational deliberation might somehow "free" us from its more negative influences.

However it would seem that there is an inherent or unavoidable paradox caused by how affect is positioned in such claims in relation to the work that making such claims—of taking up affect as an object of research or analysis itself—is presumed to perform for those of us interested in the ways in which social power or cultural conditioning takes place. As Ruth Leys concisely summarizes it, if there is one claim that unites recent critical work on affect recently in the humanities and social sciences, it is that

> we human beings are corporeal creatures imbued with subliminal affective intensities and resonances that so decisively influence or condition our political and other beliefs that we ignore those affective intensities

and resonances at our peril—not only because doing so leads us to underestimate the political harm that the deliberate manipulation of our affective lives can do but also because we will otherwise miss the potential for ethical creativity and transformation that 'technologies of the self' designed to work on our embodied beings can help bring about. (436)

As Leys goes on to suggest, there is certainly a kind of question-begging about intention taking place here—is affect the invisible force that interpolates me without my conscious knowledge, or a conduit through which I might make the conscious decision to transform myself?—that seems to repeat, at a level removed, the same "chicken or egg" conundrum that marked theories of social construction or ideological submission that affect was designed to remedy. More pressing, at least for our purposes here, is a disjunction that we might take be either broader or more specific, the contrast between how one balances what Massumi calls the "autonomy" of affect—its alterity to, and potential for freeing us from, traditional forms of social conditioning as well as of the critical analysis of culture—and the positing of affect as being at the same time central to contemporary social life.

On the one hand, affect is taken to be interesting precisely because of its overwhelming singularity or resistance to our traditional methods of quantification or description. Thus, for instance, Massumi describes affect and affective "intensity" using such terms and phrases as "unassimilable," "outside expectation," "in excess of any narrative or functional line," and as an "irreducible excess" (85–87). Similarly, affect is often positioned to be most recognizable through its disruption of the more apparent or stable qualities of cognition or cultural experience. What Massumi refers to as the "state of suspense, potentially of disruption" (86) that characterizes affective intensity is echoed by Sedgwick, who refers to affect as

> a kind of free radical that (in different people and also in different cultures) attaches to and permanently intensifies or alters the meaning of—of almost anything: a zone of the body, a sensory system, a prohibited or indeed a permitted behavior, another affect . . . a named identity, a script for interpreting other people's behavior toward oneself. (62)

Affect's autonomy from cognitive or conscious perception, or the "generalizable" in some abstract sense, is essential to its appeal as a resource for rebooting the critical analysis of culture that may have too narrowly focused on the ideological or psychoanalytic as "master" categories for studying contemporary sociality; as Claire Hemmings suggests, the most prominent work in affect theory in the humanities share an emphasis on "the unexpected, the singular, or indeed the quirky, over the generally applicable, where the latter becomes associated with the pessimism of social determinist perspectives, and the former with the hope of freedom from social constraint" (550).

On the other hand, it would also seem that the ubiquitous or even quotidian nature of affect—its saturation into contemporary culture as a whole—is equally essential to its potential for mapping contemporary culture and, in particular, emphasizing what may be new about systems of community, media, technology, and social power in the present. Even as affect names this particularly "intense" category of subjectivity-altering events and the locus of unmediated feeling within the human body, it also seems to be the very air and atmosphere of our mundane, if hypermediated, experience of life in the present.

One could, of course, dismiss this apparent contradiction as itself merely further proof of affect's ostensible centrality in human (cultural) experience: it's "both" the banal substance of our everyday experience and the charged interruption of the same; precisely because affect is something like a matrix for the formation of psychic associations and dispositions, it is also the same mechanism through which they might be altered. Perhaps, as Negri contends in another oft-cited description of the "immeasurability" of affect, "the Sublime has become the new normal" ("Value" 87). Such a conclusion may not be altogether inaccurate—and I will suggest something of the same below—but to too easily presume that affect can be a kind of privileged transformative force that might effect a more "authentic" or self-actualizing relationship to oneself, while also being the bread and butter of everyday social existence, seems to neglect a whole host of ethical and political contradictions that themselves might be taken to be lurking "just beneath the surface" of such considerations.

More specifically, if one senses a certain disjunction between these claims for affect's role in contemporary cultural domains, it might be caused not so much by its alternate positioning as either extremely rare or overwhelmingly common, but among or in between our consideration of affect as a uniquely human or least uniquely biological capacity and of its more "strategic" use as a method for motivating or manipulating individuals to perform particular actions based on their existing dispositions. As alluded to above, at least one significant driver of the recent turn to affect in contemporary criticism has been its positioning as a quintessentially human vector, an essential if often neglected component of human cognition and subjectivity that cannot be replicated in mechanical realms. Thus, for instance, N. Katherine Hayles joins many others in emphasizing the ways in which outsized claims made in the pursuit of "artificial life" or intelligence in computational systems can be debunked through emphasizing their historical disregard for the importance of affect (*Posthuman* 245–246). To give just one more example, some of the most prominent research into affect and emotional processing within political science and political psychology research has derived its central premises from Herbert Simon's emphasis on the role of emotion in human thinking, and the failure to consider it within debates over human and "mechanical" intelligence.[17]

However, subsequent research of affect has somewhat paradoxically underscored the similarities between at least some major parts of human behavior and "mechanical" processing. Take, to use an example that might play nicely off of Massumi's influential emphasis on the "autonomy" of affect, social psychologists John A. Bargh and Tanya L. Chartrand's study "The Unbearable Automaticity of Being," a consideration of the ways in which "most of daily life is driven by automatic, nonconscious mental processes" that develop from "the frequent and consistent pairing of internal responses with external events" (464, 468). Among other research, the authors review a number of studies (by themselves and others) on the effects of stereotypes on behavior in support of their thesis. For instance, subjects "primed" with words relating to stereotypes of the elderly ("Florida," "sentimental," etc.) subsequently behaved in line with the stereotype (walking slowly down hallways, having difficulty with their short-term memory). In another series of experiments, participants were subliminally presented with the faces of young African Americans; their subsequent behavior was markedly more hostile (as opposed to the control groups in the experiments), presumably based on their conceptions of that group. Given this last example, some readers might take Bargh and Chartrand as being a tad blithe in their conclusion that we should consider "automatic" affective processes as "mental butlers" acting "in our service and best interests" and who "know our preferences . . . so well that they anticipate and take care of them for us, without having to be asked" (476). More generally, however, such descriptions of the "automaticity" of affective response should give us pause over both what Hemmings codes as the "theoretical celebration of affect" in left-oriented cultural theory as well as the common positioning of affect as a key contrast to the mechanistic or computational logic of contemporary culture and technology (550). Indeed, it might be just as easy to say that affective response, particularly in its baseline logic of adaptive response and dispositional orientations, provides our most intense functional overlap with contemporary technologies.[18] In this reading, if the turn to affect is at least in part a response to anxiety over the replication of human capacities in non-human realms, or the more general integration of computational media and technology into ever more aspects of contemporary culture, it is a particularly paradoxical one: an embrace of the very qualities that are in many ways the most synthetic.

Perhaps more immediately pressing, however, is the disconnection that seems to lie between the association of affect with the most duplicitous and regressive of contemporary motivational strategies and its potential to be the basis for a revived political movement with progressive and liberatory objectives. As Hemmings details, the "optimism of affective freedom" that dominates contemporary humanistic research has often led us to neglect affective attachments with less desirable implications—"the delights of consumerism, the feelings of belonging attending fundamentalism or fascism"—introducing them only as either in contradistinction to the more intuitively "positive" affective dispositions or as the domain of actors with far less enlightened, if historically more successful, programs behind their use of affect in persuasive strategies (55).

Indeed, many humanists and social scientists with left-oriented political goals have underscored the greater acumen of political conservatives in affective political strategizing. For instance, Massumi concludes his "Autonomy of Affect" by noting that "the far right" rather than the "established left" have been far more attuned to the political potential of affect (105–106). Similarly, Lauren Berlant has somewhat gloomily argued that one of the lessons of Kerry's loss to George W. Bush in the 2004 U.S. presidential election was the importance of modeling "affective continuity" with the voting citizenship, one that was achieved much better by Bush and may be entirely unavailable to politicians that insist on foregrounding a "rational" approach to policy issues, or who prioritize complexity and nuance over certainty and "authenticity." What is often unspoken in such calls is that the appropriation of "affect-based" strategies from the right and for the left may require the adoption of practices often taken to be manipulative, deceitful, or contradictory to the values behind the objectives being forwarded—that the call for a "post-critical" or post-ideological perspective that breaks from critical theory's traditional concern with identifying what practices "enforce" or "resist" the dominant order of political economy, may require the sacrifice of a certain ethical clarity.

THE TRANSHUMAN CONDITION

Given that our concern here is with a number of phenomena that seem to short-circuit well-worn distinctions between, on the one hand, the material and the representational and, on the other, that of the natural and the artificial, we might conclude by turning to a phenomenon that might be viewed as both our best example of existing attempts to address such changes in contemporary humanities and social sciences scholarship, while at the same time as also something of a "symptom" of this process itself: the various ways in which the question of "the human" as a material being (i.e., the precise determination of what is constitutive of the human as an organism or recognizable entity) has become tied to broader questions about ethics, justice, and the intersection of the biological and the political. While perhaps the most immediately urgent exigencies for these questions emerge from practices recently only available or imaginable due to advances in technosciences, such concerns have also become central to a wide variety of policy deliberations and public conversations over a host of subjects—from human rights issues, to animal rights, to end-of-life controversies, to debates over global warming and climate change—and the focus of an immense amount of work in contemporary philosophy and cultural theory that seek to either recuperate or refract humanism as a series of precepts for working through questions of ethics, justice, and (social) responsibility.

As suggested in the introduction to this book, Julian Huxley was perhaps the first to explicitly broach this concern through his writings on the transhuman. While in the intervening decades this specific title has primarily been invoked only in specific reference to the alteration of human genetics or the human body (of which, more in a moment), over the last decade or so there have been a variety of formal attempts to engage Huxley's broader argument about the need to reinvigorate and refine the tenets of humanism in response to the demands of the present. What is perhaps most notable about such attempts is how they come closely on the heels of what often seemed to be the popular acknowledgement of the obsolescence of humanism, its eclipse as a viable ethical or political foundation beginning in the tumultuous postwar period of international politics and intellectual thought that begin shortly after the time of Huxley's writing in the mid-twentieth century. Recall, for instance, that the growing suspicion of or outright opposition to post-Enlightenment humanism's presumed reliance on a too-easy notion of shared identity (one whose attempts to include as many people as possible into a single category seemed to actually function by excluding or making abject others), and the power of human reason (one that seemed to promise much more than it could deliver) in the early works of such writers as Barthes, Derrida, and Foucault led to the popular identification of their works as inaugurating an age of "antihumanist" thought.

This term may have rather quickly fallen out of fashion but served as a crucial precursor for the more long-lasting titles of "postmodernism" and "poststructuralism," intellectual movements that were and are themselves frequently described in terms of their critique or dismissal of humanist touchstones. However, as many of the defining characteristics we use to associate with postmodernism seem to be falling by the wayside, contemporary critical thought seems to find more of a coherent identity around certain "neo-" or "post-" humanisms that draw their power not so much from some full-scale attack on humanism as a dominant cultural ideology or discourse (which is to say not so much from "negative" critique or acts of demystification in general) as they do from a more nuanced attempt to rebuild or retrofit many of its goals through less exclusionary or problematic means.

Perhaps most prominent among these discourses has been scholarship traveling under the rubric of "posthumanism," work in humanities and social sciences disciplines that show a high degree of variability in their methodologies and approaches but share a focus on critiquing the more moribund legacies of Western cultural humanism as well as attending to how research in the "hard" sciences may (or should) be altering our understanding of the constitutive qualities of human beings. While the term "posthuman" has an interesting history (including appearances in horror and "weird" fiction of the early twentieth century), its popularization in cultural criticism is largely due to N. Katherine Hayle's use of the term to describe how work in early information theory and the thematic trends of contemporary science fiction jointly present an understanding of human being and human bodies as essentially "informatic" or "virtual"—able to be "captured" by genetic code or simulated by information technology or digital media. For Hayles, such a development is both undeniably modern—made possible by a number of entirely novel developments in scientific and aesthetic thought—while at the same time an intensification of a certain idealist and abiological conception of human consciousness that finds it roots in Platonic thought, a privileging of human reason over the messier domains of human biology and affect. The posthuman then becomes for Hayles both a description of this contemporary situation as well as the possibility of configuring its tendencies differently; we might do better, Hayles suggests, in continuing the development of biotechnology and digital media while still maintaining an ethical perspective "that recognizes and celebrates human finitude as a condition of human being, and that understands human life is embedded in a material world of great complexity" (*Posthuman* 5).

More recently Carey Wolfe has provided an influential and nuanced program for posthumanism, one that expands its scope into a broader and more thorough rethinking of the work of the humanities to account for the "necessity of new theoretical paradigms . . . a new mode of thought that comes after the cultural repressions and fantasies, the philosophical protocols and evasions, of humanism as a historically specific phenomenon" (xvi). Much as humanism was a robust content provider for Western ethical thought for several centuries, Wolfe proposes an equally wide-ranging posthumanist perspective that might, somewhat more humbly, and against the temptations of anthropocentrism, provide an orientation for rethinking human identity and ethical action within a rapidly changing natural and social world.

We might identify another category within current attempts to rethink or retrofit humanistic concerns around works that are responding to much the same concerns as Hayles and Wolfe but recommend much the opposite conclusion in response. In opposition to Hayles' or Wolfe's suggestion that recent scientific and social developments call for a thorough critique of humanism as a philosophical worldview, contributors to this perspective instead argue for the need to reinvigorate or recommit to fundamental principles of Western humanism as a corrective response to these developments. What we might call an emergent "neo-humanism" of this type is well represented, for instance, by the prominent American neoconservative thinker Francis Fukuyama's work on the (present and predicted) social impact of psychopharmacology and biotechnology. Fukuyama is perhaps best known for his 1989 declaration that the end of the Cold War and the spread of Western liberal democracy and capitalism marked the "end of history." However, writing a little over a decade later, Fukuyama would revise his thesis to suggest that the process of "history" has begun again around competing struggles over the future of human nature rather than of human societies. Whereas the explicitly political conflicts over the best foundations for government certainly involved broad questions of the justice and the "good life," for Fukuyama, new techniques of altering human genetics and the popularization of neuropharmacological agents such as antidepressants prompt dilemmas over even more basic questions regarding the natural condition of the human and the ethical limits that should be placed on human enhancement. Such challenges, Fukuyama argues, require something like a radical recommitment to normative protocols of human being and behavior as well as our collective investment in the belief that "human nature exists, is a meaningful concept, and has provided a stable continuity to our experience as a species" (*Posthuman* 7). Though writing against the backdrop of a somewhat different political tradition, the German philosopher Jürgen Habermas has made strikingly similar claims in his recent work. Like Fukuyama, Habermas also contends that "new technologies make a public discourse on the right understanding of cultural forms of life in general an urgent matter" (*Future* 15). While particularly sensitive to the ways in which humanist philosophies or definitions of human nature have been deployed to the often brutal disadvantage of particular groups, he also follows Fukuyama in arguing for a generic and recognizable human nature or "anthropological self-understanding of the species" that is constitutive of humans as social beings, and which is in need of governmental protection in a time wherein it is possible to fundamentally alter the human germ line.

Finally, we might consider a more diffuse but equally influential recuperation of central tenets of humanism that Bonnie Honig has referred to as a newly apparent "tragic" or "mortalist" vintage of humanist thought. As Honig suggests, participants in this movement take their cues from the turn to ethics that became prominent in critical theory of the humanities and social sciences near the end of the twentieth century, but in many important ways look much further backward to the "tragic sensibility" of early Greek philosophy and art. More specifically, Honig writes, recent work in political theory by such writers as Judith Butler, Nicole Loraux, and Stephen K. White carries forward the critique of humanism leveled by much poststructuralist philosophy, but "reprises an earlier humanism in which what is common to humans is not rationality but the ontological fact of mortality, not the capacity to reason but vulnerability to suffering" (73). In such a "tragic" humanism, we have what appears to be something like a compromise between, or hybridization of, the viewpoints of what Wolfe codes as posthumanism and the implicit neo-humanism of thinkers like Habermas and Fukuyama: while maintaining what we might take as a foundational humanist investment in a generic conception of human nature, in "tragic" humanism these qualities are explicitly embodied and material ones. Indicative of a vision that has learned much from the failures of more "triumphant" versions of post-Enlightenment humanism, tragic humanism also trends more toward identifying a fundamental fragility or of shared vulnerabilities of the human, as opposed to the more affirmative beliefs in the mastery of reason or in human supremacy over the rest of the natural world that is traditionally associated with historical humanism.

As theories of contemporary culture, all of these approaches are in a sense descriptive—they all respond to specific stimuli novel to the present moment, even if, in the case of a writer like Fukuyama, they end up reaffirming some fairly traditional notions—while at the same time bringing their own prescriptive charge to the phenomena under their shared purview. While much could be said about the differences between how Hayles or Wolfe conceive of or define posthumanism, or about, despite their mutual investment in entertaining certain primordial character traits as quintessential to humanity, the significant distance between the more general neoconservative politics of Fukuyama as compared to Habermas' moral liberalism, I want to focus here on a specific perspective that they all seem to share. Doing so, I think, will allow us to get rather quickly to the intersection of ethical frameworks and practical concerns within these discourses, as well as to perform some useful groundclearing around this text's interest in the "transhuman" as a conceptual category.

Despite their great divergence on other issues, the various post-, neo-, and tragic humanisms of contemporary critical theory come together in their mutual disaffection for what we might consider the more triumphalist or audacious vintages of historical humanism, the various ways in which humanist thought has manifested as either a determined anthropocentrism or the more general contention that humans have the right to alter the natural world, their own bodies, or (their) nature for their own purposes. This criticism is, of course strongest in the neo-humanist opposition to genetic engineering covered above, but equally informs the posthumanist views of writers like Hayles and Wolfe. Indeed, Hayles builds her analysis of the cultural implications of contemporary scientific research precisely around a fairly pointed critique of such supposedly hubristic pursuits as radical human life-extension or the transference of uniquely human capacities into mechanical realms, goals that she reads as emblematic of liberal humanism's overbearing confidence in the autonomy of human consciousness and reason; Wolfe likewise explicitly identifies his posthumanism as one pointedly oppositional to similar pursuits and the more general contention that human beings can transcend their animal or biological nature (xxi–xvi). Similarly, while the variety of writers that Honig identifies as taking part in a new mortalist or tragic humanism have not nearly as directly engaged with contemporary developments in science and technology, as the "mortalist" designation suggests, their focus has very much been on the privileging of a "the human" in a way that cuts starkly against humanism's more affirmative or assertive stances, emphasizing the "humble" or fragile core to our shared humanity, our essential finitude and generic vulnerability to the losses and suffering caused by war, disease, disaster, and death.

It is however the moments in which these discourse directly address questions concerning the use of biotechnology or genetic engineering that the ideological coherence of contemporary attempts to rethinking humanism for the challenges of the present is most apparent. In particular, we might locate a certain shared ethical or even normative perspective around these issues, one that operates well beyond the not insignificant pragmatic concerns that might be associated with such practices. In other words, while there is certainly reason to be concerned about what we might call "Pandora problems" in reference to practices of genetic alteration—that what appear to be relatively benign procedures might open the door to a variety of unforeseen dangers—it seems that criticisms offered by contemporary (post)humanisms up until this point have largely focused on the potential for what we might alternately code "Prometheus problems"—the concern that such events will either fundamentally disrupt the "natural order of things" and/or express an overwhelming arrogance in the right and power of humans to dramatically alter themselves and their environments at the expense of other humans, other non-human animals, or the natural world as a whole.

What interests me about this particular contrast—one between the pragmatic as opposed to normative vectors of contemporary work on (post) humanism—is the ways in which it seems to repeat a certain dynamic that would seem endemic to humanist thought of any era: the ways in which any attempt to position "the human" in relation to other species or natural phenomena, even those explicitly opposed to anthropocentrism or the "autonomy" of humans from the rest of the natural world, tend to end up reinforcing this very supremacy or autonomy, the position of the human as the one who "decides" to take on such a role. In other words, it seems to me that there is something of an inherent tension produced when, for instance, Hayles or the various "tragic" humanists identified by Honig suggest that humans might undercut their more promethean tendencies by embracing their essential mortality or finitude; we find the same tension, to give just one more example, in Wolfe, writing in the context of animal rights discourse, encourages ethically minded readers to acknowledge that humans do not have sole purchase on the cognitive and emotional qualities that we associated with subjectivity, and warns that failing to acknowledge the nonhuman subjectivity of animals in particular is an example of an inchoate *speciesism* subtending even a variety of otherwise progressive thought on ethics and justice.

On the one hand, such calls urge us to interrogate or reject the more arrogant or self-congratulatory qualities of historical humanism, the too-easy conclusion that humans are the center of natural life and have the privilege to remake the work in fulfillment of their own decisions and desires.

In particular, as many of these writers are quick to mention, such presumptions of "human supremacy" or our implicit speciesism have at least significant conceptual overlaps with a wide variety of other "isms" and retrograde political and ethical positionings (including, most obviously, those that have led us to identify "other" humans as not, or at least not equally, worthy of the designation). On the other hand, however, these suggestions also seem to emphasize that very power or unique capacity that such calls are intended to undercut, the unavoidable position of "the human" as the species that gets to decide who or what can be classified in that category and the priority to be assigned other species. In this sense, then, even the most strident critiques of humanist hubris or anthropocentrism seem to end up at least implicitly reaffirming what Giorgio Agamben has recently called, after Heidegger, "the open": the essential or constitutive power of humans to decide how the natural world is divided, one that makes even our sympathetic concerns for how worthy non-human animals or other phenomena are capable of thought, feeling, or sympathy itself an appropriative projection of "the human" into the nonhuman world (*Open* 57–62).

Consider for instance, an example of this dilemma as presented by Wolfe, the writer who perhaps more than any other has attempted to address the pragmatic challenges of thinking "beyond" humanism. Writing in regard to animal rights, Wolfe presents a qualified endorsement of the Great Apes Project—a movement to secure a United Nations guarantee of certain basic legal rights for non-human hominids (chimpanzees, gorillas, and orangutans)—in a way that expresses sympathy for its mission, but concern about the ways in which the campaign leverages the genetic similarities between humans and the other Great Apes in its messaging and public awareness campaign. As one might expect, given that humans are also one of the species designated with the category of Great Apes, promotional material for the campaign has drawn heavily on appealing to potential proponents and campaign contributors by emphasizing the "human" qualities of these animals. While such calls seems to have much in common with Wolfe's interest in combating the historical anthropocentrism of Western ethics, he cautions that "the model of rights being invoked here for extension of those who are (symptomatically) 'most like us' only ends up reinforcing the very humanism that seems to be the problem in the first place" (*Animal* 192). We come to sympathize and support protections for the Great Apes not out of respect for their alterity or difference from us, but the ways in which they appear similar to humans as opposed to other, "lesser" animals, and thus even an animal-rights discourse such as this one ends up reproducing the logic of anthropocentric humanist ethics, simply one step removed.

However, allow me to turn the screw one more notch and suggest that, despite the inherent problems that occur when we depend on shared (human) qualities as a basis for ethical responsibility, it is difficult to imagine any kind of ethics that does not rely on such a strategy of demarcation and communion, or, perhaps more problematically, that we encounter the same problem in reverse whenever we try to consider an ethics not based on shared qualities or capacities. Indeed, shortly after his discussion of the Great Apes Project, Wolfe consider a scenario somewhat like this in wondering whether there is a similar contradiction in humanist ethical thinking that judges the quality of ethical action around its degree of "selflessness," the extent to which it is preformed without any guarantee or presumption of a personal gain or reciprocal action in kind from its recipient. By these standards, Wolfe suggests, would not the "supremely moral act" then be directed not toward another human who one could at least imagine as repaying it in some way, but rather toward "the animal other, from who there is no hope, ever, of reciprocity?" (*Animal* 199). While Wolfe goes on to imply that there is a certain problematic moralism in such thinking as a whole, we might seize on such a comparison to think about the possible "other ends" of the reluctance to base ethics on shared identity or capacities. There are, of course, additional categories other than the non-human animal that would stand even less of a chance of reciprocating our ethical behavior, or of appealing to our sense of shared identity: plants, machines, inanimate objects of all types. While it might be absurd to imagine ethical action devoted to, for instance, a rock, such a comparison is perhaps useful for thinking through what we might take as the parallel dilemma to ethical systems that demand some "human" qualities in order to be considered worthy of ethical consideration. Far from being an extraordinarily selfless, disinterested, or "nonhumanist" act, trying to act ethically in relation to a rock would only reveal how such activity was always about the doer herself, how the only objective involved in the undertaking was the "performance" of ethics itself, one without any pragmatic goal beyond a certain self-satisfaction. Taken together these two extremes form the bookends of what we might call "the humanist paradox": any ethical system based on the recognition of shared qualities threatens to exclude others through a certain myopic series of self-identifications, but any attempt to eschew this procedure entirely *also results* in the revelation of a fundamental self-interested or solipsistic relationship to one's entire environment. Once could, of course, expand this paradox into a number of related directions, arguing, for instance, that calls in contemporary cultural and political theory that counter more promethean versions of humanist thought by urging us to embrace the essential "finitude" of humans, or those that urge us to increase our affective or communal ties around our shared vulnerability operate via similar logic: as strategies that leverage their audience's existing pretensions to a ethical or intellectual superiority under the guise of asking them to renounce or see beyond these traditionally "humanist" stances.

My interest here, of course, is not in suggesting that this tension or conflict leads to some kind of unavoidable contradiction within the various arguments made on behalf of new work on (post)humanism *or* in campaigns such as that of the Great Apes Project, not to mention the variety of human rights discourses, that explicitly invoke a "shared humanity" to in their arguments and appeals—indeed, I will end up suggesting that this is precisely what makes them persuasive and powerful. I am, however, for the remainder of this chapter, interested in thinking through a related question about the past and future of humanist thought. Specifically, I am interested in considering whether the turn to a more "materialist" version or critique of humanism—one that emphasizes the embodied and biological components of human experience and thought—has concomitantly emphasized what we might call the more "material" effects of humanism as a philosophy or code of ethics, what we might call the matter of what humanism *does* or the *effects it has produced* as opposed to what it has *meant* in its various historical variations, the ways in which humanist discourse "self-presents" or explicitly details the contents of its own thinking.

The antihumanist discourse of postmodern and poststructuralist thought certainly took humanist philosophy to task for its various contradictions or shortsighted perspectives—the ways in which particular conceptions of ostensibly essential and transhistorical human characteristics seemed significantly influenced by particularly cultural contexts, or how the humanist faith in human reason seemed to involve a denial of the role of social power in human cognition—such salvos were typically designed to show the superiority of a different theoretical perspective on human culture, or to encourage a more abstract skepticism of dominant modes of social thought; insofar as humanism was taken to be something like a codeword for the various ways in which normative behaviors authorized by structures of social power had come to take on the appearance of being entirely "natural," critiques of humanism found their power precisely through deflating the apparently incontestable commonplaces that had developed around such constructions. More recent engagements with humanism's legacies in contemporary cultural theory, however, seem to be after somewhat different quarry, particularly in regard to their relationship to normativity or what we might generally refer to as the prescriptive rather than purely diagnostic mode of criticism (the vector that has given much recent work in the critical theory wings of the humanities and social sciences its ethical character). More precisely, we might say that what distinguishes recent work in cultural and ethical theory from more recognizably postmodern or poststructuralist critiques of humanism is the ways in which these newer discourses have, on the one hand, extended the target of critique (from a retrospectively narrow focus on social conventions and institutions to one that includes questions of human embodiment and biology) while, on the other hand, still seeking to preserve the normative power that long associated with humanistic thought, but one now to be deployed in more egalitarian or socially ameliorative ways. If there is a tension between these two objectives, it is perhaps most apparent in the paradox mentioned above: any attempt to convince individuals to move "beyond" or "outside" of an anthropocentric or self-derived series of identifications with others seems to only be possible via a kind of implicit or even underhanded invocation of the same strategy.

Rather than contending that this paradox inherently short-circuits such attempts, however, I want to shift ground a bit and suggest instead that perhaps it is this very strategy—rather than any particular epistemological perspective or ideological content—that has long been the constitutive core of humanist thought throughout its entire history. In other words, the various contradictions that seem to rise immediately in attempts to either rethink or reject historical humanism may have not so much proven that it is impossible to abandon the enterprise—though this may also be the case—as it has revealed what has given it a coherent identity despite having taken a variety of forms in its several-centuries long history. Perhaps this suggestion—that we might have to reconsider humanism as something of a rhetorical strategy rather than a philosophical doctrine—is best unpacked via a distinction Foucault made several decades ago in his essay "What Is Enlightenment?" While the popularization of poststructuralist and postmodern theory at the time during when Foucault is composing this essay tended to lump the implicit antihumanist tone of these writings alongside their more specific critiques of Enlightenment thought and post-Kantian philosophy in particular, Foucault here cautions us to avoid the "always too facile confusions" between these two (315). This confusion, Foucault goes on to suggest, is the result of not only the complex nature of both items under review, but also of our desire to in some way purify or modify our conceptions of them in order to more easily endorse or more fully reject them. Thus, as Foucault writes, one way people try to avoid what he calls the "the intellectual blackmail of being 'for or against the Enlightenment'"—opposing or embracing it in the abstract—is to introduce the "dialectical nuances" of determining its good and bad qualities as if they could be separated and judged individually (314). In what would seem like a fairly clear gesture toward attempting to escape such a forced choice, Foucault somewhat famously associates Kant's response to the titular question ("what is Enlightenment?") with a project of performing "a permanent critique of ourselves" that he will propose for contemporary philosophical thought, even if his earlier work was often read a condemnation of sorts Enlightenment discourses rationality, and of Kant's contributions, in particular.

Foucault's engagement with humanism is much briefer, even as Foucault seems to be suggesting it is the more vexed of the two concepts. Whereas the Enlightenment can at least be grasped as "an event, or set of events and complex historical processes, that is located at a certain point in the development of European societies," humanism, Foucault writes, "is something entirely different . . . It is a theme or, rather, a set of themes that have reappeared on several occasions, over time, in European societies; these themes, always tied to value judgments have obviously varied greatly in their content as well as in the values the have preserved" (313). Foucault goes on to demonstrate this variety in "content" and "values" by mentioning that the seventeenth century witnessed a humanism premised on a critique of (Christian) religion, as well as a Christian humanism, and in the nineteenth century, humanisms both premised on and opposed to scientific thought, and later, humanisms that were Marxist and existentialist, a humanism aligned with National Socialism, and one claimed by Stalinists. From this trajectory Foucault concludes "the humanist thematic is in itself too supple, too diverse, too inconsistent to serve as an axis for reflection" (314).

Written against the backdrop of an upsurge in populist political movements, and the even more specific contexts of the rather pitched "theory wars" of the mid-1980s, Foucault's "What Is Enlightenment?" is often read as, if not a manifesto, then at least a significant groundclearing exercise. If the by then well-worn critique of postmodern or poststructuralist thought was that it seemed to be a largely negative or even parasitic movement—one that made its case by exposing the hypocrisies or inconsistencies of dominant beliefs or of institutionalized knowledge, and often via what at least appeared to be a critique of "reason" or of human agency *in toto*—then Foucault's intervention in "Enlightenment" was to illustrate how one could carry on the work of critique while still staying true to the progressive principles of the Enlightenment. Foucault argues that the urge to undergo a "critique of ourselves" is a not altogether unfamiliar update of the Enlightenment's ostensible goals of banishing superstition and reforming society— and this goal was itself not necessarily contradictory to a more dynamic, but still recognizable concept of human agency; the two are intertwined, in Foucault's word in "the principle of a critique and a permanent creation of ourselves in our autonomy" that he offers as a modern-day inheritance of the historical Enlightenment (314). In other words, and to put things a little more pointedly, what Foucault accomplishes in this piece is a particular refinement, one might even say cleansing, of the school of late twentieth-century critical thought of which Foucault himself was an important originator, one that defends poststructuralism from any lockstep portrayals of it as "antirational" or purely de(con)structive. In doing so he issues a fairly strong endorsement of Enlightenment thought of at least a certain type, that complex of directives and objectives that was presumably the key target of postmodernism or poststructuralism and of which critics of these later movements sought to protect, while simultaneously attributing or transferring all of its exclusionary, imperialist, dogmatic or otherwise negative qualities to "humanism," that phenomenon that Foucault tells us has always been consistently confused with the Enlightenment.

Given that Foucault took his own investigation into the histories of these terms to also, and more importantly, be a question about the present, specifically a way of repeating a question that he sees implicit in Kant's essay on Enlightenment—"What difference does today introduce with respect to yesterday?"—what can we make of Foucault's distinction today, over two decades since the composition of "What Is Enlightenment?" (305). What is the status of these two complexes—one, "Enlightenment," based on the "principle of critique and the permanent creation of ourselves in our autonomy," and one, humanism, that finds its identity via reliance on notions of human constitution—in a time wherein an emphasis on critique and autonomy seem to be the basis of social, political, and economic discourses of all stripes, including the most retrogressive or conformist, and during which the "constitutive" elements of the human seem to be under the permanent threat of dissolution via their appearance in nonhuman realms?

Perhaps the answer, much like Foucault's own rereading of Kant's work as a reflection on the histories of Enlightenment and humanism, might be made apparent through a rereading of Foucault's own "updating" here, one informed by the other cultural changes engaged in this chapter. As we have seen, a wide range of novel phenomena occurring since the time of Foucault's writing might be identified around the rearrangement of our traditional divisions between the representational or ideational and the technical or purely strategic. This is the thread that unites a capitalism increasingly less dependent on the maintenance of ideological categories and cultural conformity, an understanding of human affective response that shows it to be the aspect of our experience that has the biggest correspondence with the behavior of animals and machines, and a new relationship between (scientific) knowledge and social power in which direct application and intervention crowds out the more traditional objectives of setting the boundaries of knowledge and objectivity.

Against this backdrop it is perhaps easier to perceive a different relationship between Enlightenment and humanism, or perhaps, to follow Foucault's gesture, understand something about the latter's historical identity that was up until now only obscure. If shifting grounds in what counted as knowledge and as autonomy in the time of Foucault's writing helped expose what was "essential" to Enlightenment and showed humanism to be "too supple and diverse" to have its own identity, the recent changes in technoscience and culture leading up to today—those that have shown "the human" to be itself a supple and diverse category—suggest a different distinction. It is not the construction of "knowledge" or "truths" that seem, *pace* Foucault's earlier inquiries, to be the major frameworks for social power in the present. Instead, it seems that political economy and contemporary culture as a whole is threaded by a logic of *pure technique*, or series of strategies for flexibly organizing and manipulating a variety of generic dispositions and inclinations present in the social field at any given time. In this sense, perhaps it is now humanism, particularly the strategies of the kind of "triumphant" humanism that has been under critique very much of late, rather than Enlightenment that becomes a particularly privileged discourse or pragmatic tool for responding to such a situation. After all, is it not humanism, as both its critics as well as the more material changes in contemporary technoscience and culture make clear, that has always been our name for the particularly strategy in which an individual's extant dispositions, particularly their most lockstep and self-interested ones, are directed toward or bundled together with particular goals or objectives?

This is, I take it, a quality visible in humanist discourse as early as the declaration that "man is the measure of all things," the statement attributed to the Greek sophist Protagoras that is often positioned as the earliest statement of the humanist ideal. Because this statement became newly prominent among mid-twentieth-century epistemological debates for its supposed endorsement of cultural or subjective relativism, we have perhaps missed the ways in which it also presumes a certain, necessarily self-interested, responsibility on any one that would take it as truth. As Isabelle Stengers suggests, it offers the mantle of being the one who "measures" as "a requirement and not a destiny," a necessary obligation rather than an essential right (*Invention* 165). In doing so, such a declaration is both a description of a practical obligation and a fundamentally pragmatic endeavor on behalf of its author, one that, as Stengers suggests, also reveals, in the sophist's thinking, a desire to interrogate the breakdown between "human affairs (*praxis*) and the management-production of things (*techne*)" that were being divided in early Greek philosophy and politics. Far from being its own act of (counter-)Enlightenment—the revelation that all systems of measurement and verdicity are a social construction—it instead functions as a rhetorical strategy for bundling the more solipsistic or self-entitled tendencies of individuals with a particularly set of responsibilities. Foucault's conception of humanism as a phenomenon that is unable to be its own "axis of reflection" is still (has always been?) accurate; rather, humanism has been a particular rhetorical strategy, but one whose functioning and component parts have perhaps only recently been made clear as the kind of veridical structures of identity often taken to be humanism's core have become less crucial to the operations of social power.

Similarly, to suggest a reorienting of priorities that might extend Foucault's gesture at another level, perhaps what these teaches us is that it is not the "permanent critique of ourselves" or the "creation of autonomy," nor, for that matter the kind of "critique of our autonomy" that seems to drive at least much posthumanist discourse, that seems to be the most urgent or potentially useful endeavor for political and ethical thought in the present. Rather, it is likely that identification and mastery of these kind of strategic techniques focused on producing actions rather than altering identities, and that work with rather than against existing predispositions, techniques that have perhaps often been obscured beneath the name "humanism," and the coming-to-consciousness of Huxley referred to as "tranhumanism."

TOWARD A RHETORIC OF THE TRANSHUMAN

So, after all this, what does this attempt at periodization, this critical reading of the cultural effects of technoscience and contemporary media over the past several decades, offer us other than a mere description of the present? In a certain sense, the revelation that the central forces structuring whatever we might call culture and social power today—from the practical foundations of contemporary technoscience, to the operations of contemporary capitalism, to the everyday experience of "being" in global mediated culture—have becoming increasingly less premised on epistemological or ideational vectors and more on the blunt application of particular generic techniques and strategies would seem to make the act of "description" itself one with diminishing returns. To compare it to one of the dialectical pairings we have had occasion to discuss here, we might say that such an analysis offers the comforts of neither humanism, as traditionally conceived, nor the "antihumanist" tendencies of postmodern and poststructuralist analyses of cultural and social life. If we take humanism, in either its vulgar sense as a essentialist and anthropocentric designation of the "proper" center of judgment and power, or in its more novel vintages based on the critique of such essentialisms and the management of human's "tragic" or "moral" nature, to be, as I have suggested here, more properly a rhetorical strategy than a epistemological foundation, then it seems to lose its normative power.

Similarly, if we take the opposite end of this dynamic—the more vulgarly postmodern contestation of such foundations—we also seem to be left with rather weak tools for responding to dominant modes of contemporary culture; as all of the phenomenon detailed above seem to testify, at least the generic forms of critique, skepticism, or "resistance" have appeared to become basic operating principles of the institutions and discourses they were designed to combat. The situation might seem even more distant from that other pairing used as a point of comparison in this chapter, the periodization of postmodernism as a cultural or historical moment distinct from the modern. After all, the key point that seemed to unite the disparate ways in which the postmodern was theorized in the latter half of the twentieth century was the general revelation that we were living in a time in which the representational, communicative, or otherwise "cultural" seemed to be emerging as the dominant logic of the socius, a relationship that seems to have been inverted in the present as technicity seems to become dominant, or become crucially mixed with the representational.

However, if we take this emphasis on technicity, the becoming-technical of culture and social power, on its own terms, we might find there is plenty of value in, and much work to do around, this kind of descriptive work or periodization. In other words, it would seem that a description of strategies is precisely what one would need to account for and respond to the present, as long as it is one that does not put its trust in the power of demystification or of new modes of representation, but dedicated toward assaying these strategies themselves *as techniques* with material effects, and seeks to find some other way to deploy them. To unpack this by way of comparison one more time, we might compare such a move with the methodologies proposed in two of the more canonical periodizations of postmodern culture.

For Jameson, the highly mediated and seemingly vertiginous reality of post-modern culture was one that might be best met by what he called a "cognitive mapping" providing "the individual subject with some new heightened sense of its place in the global system" (*Postmodernism* 54). For Lyotard, the recognition that postmodern culture was subtended by language games and shifting systems of social legitimation led him to suggest that our new objective was "not to supply reality but to invent allusions to the conceivable which cannot be presented," a mission that he famously described as a charge to "wage a war on totality" and "be witness to the unpresentable" ("What" 80).

The first distinction we might make in reference to the demands of the present is that we are not so much in need of "mappings" or new systems of representation today (or critiques of old ones), but in something more akin to a "user's manual," a pragmatic description of the various strategies and techniques available for intervening in the functions of social power. Second, we might suggest, to repeat a point already made multiple times in this chapter, our key domain for thinking through such a mission is not likely to be that of epistemology or representation, but that of rhetoric and persuasion. As I will be suggesting in the following pages, it is this vector—the marshalling of dispositions in alliance with particular actions or goals—that seems to encompass the largest share of contemporary fields of politics, affect, and ethics in the present.

In the end, then, what we might learn from an analysis of present culture that emphasizes its increasing technicity, is that the very technical or technique-driven nature of contemporary social power suggest that the tools for effectively responding to, or redirecting, its functions are all around us and available for purposes very different from the ones currently dominating present formations. It is with that objective in mind that the rest of this book proceeds. After offering a genealogical analysis of the genealogy of technics and media that led up to the present condition, it then turns to examining how the work of cultural theory might be performed in this context, the complex sets of relations between affect and persuasion in the age of ubiquitous technology, the question of "resistance" in a moment when contemporary institutions of social power appear to run on opposition, and the possibilities for ethics in a world where all "value" seems to be immediately available as a capitalist commodity.

2 The Age of the World Program
The Convergence of Technics and Media

SPIEGEL: And now what or who takes the place of philosophy?
Heidegger: Cybernetics.

> —Martin Heidegger, "'Only a God Can Save
> Us Now': The *Speigel* Interview" (108)

These puzzles where one is asked to separate rigid bodies are in a way like the "puzzle" of trying to undo a tangle, or more generally, trying to turn one knot into another without cutting the string.

> —Alan Turing, "Solvable and Unsolvable Problems" (181)

It might be said that a number of the central contentions of this book—notably that the impact of technology on culture over the past half-century or so goes far beyond the actual "use" of technology, and that the best way to frame such changes is to in some way retrace the separation of technicity from knowledge and language in early Greek culture—are largely a continuation of Martin Heidegger's writings on technology. Indeed, one might say that these concerns became the dominant themes of Heidegger's work, beginning from at least as early as the 1938 essay "The Age of the World Picture" and continuing into the 1950s wherein the "threat" posed by technology to a number of cultural processes becomes a central part of Heidegger's reflections on the possible futures of democracy, philosophy, and of "humankind" as a whole. Although critiques of Heidegger's work on these subjects have often positioned his conflation of material technologies with conceptual or interpretive frameworks—that there is something like a "technological worldview" that is in excess of our material use of technologies—as a weakness, we might today instead see this as its strength: Heidegger's insistence that "the essence of technology" transcends its "mere external forms" to radically reorganize capacities for human communication, the synchronicity or co-temporality of global culture, and humans' ability to destroy their natural environment can look like quite a prescient one when viewed retrospectively from an era marked by the ubiquity of "information technology," economic and cultural globalization, and an increasing dread over the ecological devastation caused by human industry ("What" 112).

There are, however, at least two other reasons we might be suspicious that Heidegger's critique of technology or technics, important as they may be as an anticipation of the contemporary challenges posed by these phenomena, might not be the most useful tools for responding to them today. Perhaps most obviously, the fundamental technologies of Heidegger's day are hardly the same as those that are presently at the center of contemporary culture. Although one might presume, simply by virtue of the speed of technological change since the mid-twentieth century, that any of his references to specific technologies might unavoidably seem quaint when read decades after the fact, what is far more problematic are the ways in which such changes have largely diverged from or even inverted the tendencies that were central to Heidegger's analysis. Thus, for instance, while it might have already seemed somewhat of a stretch to identify, as Heidegger does, the ability for people across Europe to simultaneously listen to a concert broadcast in London as exemplifying the ontological condition of mid-twentieth-century individuals—the "frenzy for nearness" [*eine Tollheit auf Nähe*] that drives one's interest in the broadcast evoking, for Heidegger, *Dasein*'s fear of confronting its own finitude and own "time" by embracing the generic temporality of a worldwide community—it is, of course, very difficult to imagine how this would match up with the trend in contemporary media toward asynchronicity, niche audiences, and a general "individualization" of time and temporality on the level of (media) consumption (*History* 312). While it would be hard to argue the reverse, that the increasing asynchronicity of media now stages a kind of *confrontation with* one's own "finitude" or individual temporality—there are perhaps a number of subjective effects produce by watching an endless stream of YouTube videos or of answering work e-mails in the middle of the night, but these hardly seem to be likely ones—there are, as I will attempt to show below, a wide number of constitutive divergences between the kinds of media and their concomitant representational tendencies referenced by Heidegger and those that are common today. To anticipate a bit, though, I will suggest here that Heidegger's own reference to "the new fundamental science that is called cybernetics" as the culmination of a long historical sequence of both metaphysical thought and of technical evolution more or less suggests the same: the intervening decades have very much proven that cybernetics is perhaps better grasped as the start of a radically new stage in these processes, rather than the predictable "completion" of others ("End" 58).

Second, even if we discount criticism of Heidegger's conflation of material technologies and hermeneutic structures, we might still stay that his approach to unpacking the "meaning" or "essence" of technology is limited by his more specific suturing of this investigation to the history of Western metaphysical reason as a whole. Recall that, for Heidegger, the interrogation of technology is very much bound up with a parallel, if broader, investigation into "historical ontology," the various ways in which humans have attempted at to ground their access and understanding of phenomena of any type at the most basic level. As he tells us early in "Age of the World Picture," "metaphysics grounds an age, in that through a specific interpretation of what is and through a specific comprehension of truth it gives to that age the basis upon which it is essentially formed" (115). Heidegger's suspicion of the various ways in which such a metaphysical grounding tends to identify a central force as the guarantor of truth is largely responsible for the ambivalent positioning of metaphysics in his work. On the one hand, metaphysics is shown to be much more than an esoteric concern of interest only to philosophers. Though the fairly elite discourse of philosophy remains the best resource for accessing the ontological source positioned as central in a given historical moment, it represents a much broader cultural understanding of modes of experience and self-reflection; as Iain Thomson writes, for Heidegger, "by giving shape to our historical understanding of 'what *is*,' metaphysics determines the most basic presuppositions of what *anything* is, including ourselves" (8). On the other hand, while this would presumably give metaphysics a fairly obvious pride of place in any attempt to consider dominant modes of perceiving and thinking, Heidegger's tends to read the history of what most would consider the entirety of the Western metaphysical tradition—from Plato to the philosophy of Heidegger's own contemporaries—as largely a series of compound errors.

Thus, while Plato is typically invoked as genesis figure for metaphysical questioning, Heidegger instead reads Platonic thought as the start of a gradual forgetting of the original question "of Being," the start of a long process by which one or another specific force ("ideas" of "ideal forms" for Plato, "God" for medieval philosophy) is offered to answer the most essential question of philosophical thought—"why are there beings at all instead of nothing?"—until the question itself is largely forgotten, or made a purely subjective one, insofar as human beings are positioned as grounding their own access to phenomenal reality. In this sense, then, while metaphysics is the name given to the long history of establishing the grounds of human perception and understanding, it is at the same time the history of growing further away from both the question of and our access to these categories as they were experienced by the pre-Socratics.

It is at the end of this kind of regressive sequence or history of decline that Heidegger places the "essence" of technology. By invoking the dual meanings of various terms designating the "end" or "conclusion" of something, Heidegger identifies the technoscientific domain of cybernetics as the final stage of this process: both the end of philosophy or metaphysics as the several-centuries-long attempt to answer the fundamental questions of human existence as well as the "fulfillment" or highest stage of the degraded path that this effort has taken from Plato onward. To quickly summarize a fairly complex argument that we will have the opportunity to unpack in more detail below, for Heidegger the "world picture" offered by technoscience shows an absolute disregard for questions of purpose, direction, and phenomenological or veridical grounding, and thus serves as the full or quotidian emergence of a certain kind of operational nihilism, or absolute subjectivity of experience, that was only implicit in the "anti-metaphysical" thinking of figures like Marx and Nietzsche. The essence of technology, then, appears to us as the final triumph of a purely "calculative" mode of thinking and social interaction over the kind of "meditative thinking" that Heidegger considers as historically constituting "man's essential nature" (*Discourse* 56). Technology thus becomes not only a symptom of the new phase in "human nature" but also something like a compensatory mechanism for the same cultural transformations it has helped midwife; technology, Heidegger tells us, is not only the "essence" that organizes contemporary experience as well the physical matter of a wide variety of the world, but also "the organization of a lack," a way of "keeping things moving," in a world when purpose or a sense of destination or completion seem like antiquated concepts ("Overcoming" 107). Thus, to mention only one of Heidegger's examples, socioeconomic exchange continues apace, but only via "circularity of consumption for the sake of consumption" rather than as part of some process or a larger meaning or identity.

And, I take it, it is these kinds of conclusions that are likely very familiar to contemporary ears. Though they are certainly not typically phrased in the terms of Heidegger's fairly rarified language, it is only a small step from Heidegger's positing of contemporary technology as a kind of simultaneous "shallowing" and "speeding up" of contemporary culture and our default conceptions of the social impact of communicational media and information technology today. Indeed, it could be claimed that Heidegger's writings on technology are immensely influential if largely unacknowledged sources for a wide variety of both academic and popular critiques of the effects of technology on contemporary culture, from the contention that it encourages a kind of heedless and unreflective participation in some of the worst tendencies of contemporary capitalist consumerism, to the feeling that our connections to "reality" and each other are increasingly mediated through impersonal and informatic exchanges of various types.

There is much to be said for such critiques, and I do not necessarily disagree with many of them. It would be immensely difficult, for instance, to suggest that contemporary communication media and information technology encourages the kind of "slow" or meditative thinking prized by Heidegger (and, more broadly, a certain kind of cultural humanism at large). What I am interested in, however, is troubling the frame of reference that makes such comparisons seem inevitable, as well as the not unrelated question of what that comparison, and other possible frames of reference, might mean for us today, particularly in the field that we might take as having the largest overlap with what Heidegger codes as philosophy—and the one in which he has had his most outsize posthumous impact: contemporary critical theory in the humanities and social sciences. While it may be true that "meditative thought" has been on the downswing over the last several decades—which, to be fair, would only be the continuation of a process in Heidegger's account that has been in something of a downward spiral since the fourth century BCE—it does not seem to be me like the "calculative" or similar descriptions of the inhuman or purely ends-driven logic of political economy remains dominant today; indeed it would seem that if such a regime ever did exist, it would be more accurate to locate it in the mid-century context in which Heidegger began his critique of technology, that moment in which it at least appeared that that the rolling yield of the assembly line and crossed columns of statistical tables threatened to capture us all, making us either the victims of or the mantle-holders for the "organization man" or "man in the grey flannel suit" that we heard so much about in the 1950s. All of which is to say: although I think it is certainly true that whatever we might call the dominant logic of culture today is fundamentally shaped by contemporary technology, it is hardly one that has assumed principles of calculation, reductionism, or standardization that Heidegger among many others has associated with the increased ubiquity of technology in social life.

However, while I take it that Heidegger's description of sociotechnics is largely a moribund one, I do think we might have much to learn from repeating or updating his historical take on the evolution of technology and ontology, and I will perform such a reading in this chapter, with a few key differences. More specifically, I am interested, on the one hand, in refracting or skewing Heidegger's indexing of the growing centrality of technology in social life to the decline of metaphysics from Plato onward, and instead suggest that we might more productively read this historical sequence as evincing the "return" of elements of rhetorical thought and praxis that were largely crowded out by Platonic thought. On the other hand, I am ultimately not so much interested in this particular redrawing of the boundaries or exchanges between particular disciplines as I am in the ways in which such a reconsideration might offer us a different attitude toward and concretes strategies for responding to the "technologics" of today.

While the next three chapters of this book will take up these strategies around more specific questions relating to politics or social power, the contemporary status of persuasion or motivation, and ethics, I think that in addition to doing some important groundclearing in advance of these other questions, thematizing or theorizing the dominant cultural intersection of technics and media will have much to teach us in and of itself, including some insight into the status and function of "theory" in the past and present.

HEIDEGGERIAN DETOUR: HISTORICAL ONTOLOGY AND THE HISTORY OF TECHNIQUES

But before we begin, a brief detour through Heidegger's approach to historical analysis; insofar as this chapter attempts a certain kind of rereading and response to Heidegger's project and what I take to be its influences on contemporary thinking about technology and culture, it may be necessary to provide at least a little more detail, however briefly, about that process and these effects in addition to what was sketched above. The first attempt to thematize "historical ontology"—to trace the ways that human conceptual processes and patterns of self-understanding have changed over time—is likely that undertaken by Hegel in his *Lectures on the History of Philosophy*. Early in his introduction to that work, Hegel emphasizes the apparent "inner contradiction" of that text's titular objective. "Philosophy," Hegel writes, "aims at understanding what is unchangeable, eternal, in and of itself" but "history tells us of that which has at one time existed at another time has vanished, having been expelled by something else" (7–8). Thus though one might produce something like a chronological rendering of the different propositions or methodological systems developed by philosophers, it would not seem to have the same temporal or developmental import we usually attribute to historical analysis. Hegel's unique remedy to this situation would be to combine the transhistorical or universal characteristic of philosophy with a conception of its maturation as being similar to that of an individual's learning process. Thus Western philosophy's development could be thought of as like that of "a universal Mind [*geist*] presenting itself in the history of the world"; philosophy can be traced progressively while maintaining its "universal" quality, as long as one considers its advances as proceeding in the same way that self-consciousness and maturation occurs in individual minds, but in this case distributed across different civilizations and communities (33). Hegel would further approach this question from the opposite angle in the undertaking that, appropriately enough, transposes the subject and domain of analysis of *The History of Philosophy*; he writes in *The Introduction to the Philosophy of History* that philosophy allows us to chart the coming to self-consciousness of the universal *geist* which is itself the self-consciousness of human freedom, and thus "world history is the progress in the consciousness of freedom" (22).

Perhaps what is most striking about Heidegger's particular return to this question is the way in which his own execution of a "philosophy of history" or historical ontology largely inverts both the premise and conclusion of Hegel's attempt; if Hegel's histories map a slow progression toward greater awareness and freedom, Heidegger's narrates the increasing abstraction or overmediation of human self-understanding and experience, culminating in a kind of "freedom" that is only possible in the absence of human purpose and responsibility in general. Thus Heidegger's division and analysis of historical ontology via five primary "epochs"—the pre-Socratic, Platonic, medieval, modern, and late-modern (Heidegger's present) stages of ontological self-understanding—is one in which the possibility of an unmediated and positive relation to ourselves and to "reality," one that seems at least implicit in the writings of the pre-Socratics, becomes increasingly distant as Being becomes successively grounded around the central forces of the Platonic forms, God (in the medieval epoch), the human subject (modern), and the spirit of technology or pure operationality (the present). If the sequence leading from the Platonic to the present is largely a series of errors in a too-narrow and too-easy grounding of Being that culminates in a problematic "self-grounding," the contemporary epoch that Heidegger associates with the spirit of technology expresses something like the total abandonment of the enterprise altogether. Technoscience, Heidegger tells us, takes over the traditional task of philosophy, that of determining "the ontologies of the various regions of beings," and retools it, focusing only on superficially theorizing "the necessary structural concepts of the coordinated areas of investigation," an examination of merely those structural vectors that have an apparent operational impact on function or performance ("End" 58). Thus, Heidegger concludes, "'Theory' means now: supposition of the categories which are allowed only a cybernetical function, but denied any ontological meaning."

For Heidegger, this bleeding over of the practical or utilitarian vectors of technology or technics into the realm of *episteme*, the former space for reflecting on such processes, subsequently results in a certain kind of "technologiziation" of humans themselves. While Heidegger could find recuperative tendencies in even Nietzsche's "will to power"—this reference point for Heidegger's modern "epoch" is still suggestive of a certain transformational purpose for individuals—what Heidegger calls the "will to will" that dominates human existence in the age of high technology seems to have little to no redemptive potential. Rather than a will that drives one toward the goal or the fulfillment of a particular process, even one that is destructive or self-defeating, a will *to or for something*, Heidegger suggests the contemporary age is dominated by what we might call a *will to or for anything*, an unguided and chaotic acting out of contemporary cultural forms without any particular end other than the continuation of the existing process. Thus, while the essence of technology produces what Heidegger calls an "unconditional uniformity" of humankind, it does not do so around the insistence of a particular identity or ontological "grounding," but rather via the absence of stable identities and the retreat of any ontological foundation for human existence. As Heidegger concisely summarizes this sequence in the essay "Overcoming Metaphysics," "The unconditional uniformity of all kinds of humanity of the earth under the rule of the will to will makes clear the meaninglessness of human action which has been posited absolutely" (110). Rather than being cathected to an overly rigid or narrow arbiter of truth and self-understanding, modern individuals remain in a kind of existential holding pattern, a frantic and aleatory movement within a determined space or system.

I am not so much interested in contesting the accuracy of what Heidegger identifies here as the challenges posed, or negative effects caused, by the movement of technology or technics to the center of the cultural field; indeed, as I will suggest below, there are many reasons to believe that they have become, if anything, even more urgent or "dangerous" given changes in technology and media since the time of Heidegger's writings. I am, however, interested in contesting the modes of rectification or amelioration promoted by Heidegger in response to these challenges. In other words, it is not really the *problems* foregrounded by Heidegger in his reading of technology that seem necessarily moribund from the view of present as much as it the *solutions* proposed for them, ones that I take to be not only fairly representative of contemporary thinking about technology but, as we shall see, of a wide variety of both popular and academic social theory today.

"Solutions," may be too strong of a word, given Heidegger's preference in his late work for a kind of hesitant and provisional response to contemporary culture—indeed in many ways it is "more meditation" that itself appears to be Heidegger's most common suggestion for remedying the negative social effects of the will to will, or the culture of nihilism growing around the dominance of calculative reason. What we might, however, at least call Heidegger's "strategies" might be divided into three categories. First, there is an attempt to recover the relation to Being that Heidegger sees in pre-Platonic thought, a common concern of Heidegger's late writings. The suggestion that only a "god can save us" in the title of the source for our epigraph above is, for example, a reference to such a possible new grounding; elsewhere Heidegger holds out hope that we might "stand and endure in a world of technology without being imperiled by it" through retrieving our "rootedness" or seeking "a new autochthony which someday even might be able to recapture the old and now rapidly disappearing autochthony in a changed form" (*Discourse* 55). Second, there is the very process of retracing the history of error in metaphysical reasoning that Heidegger spends a large amount of his late work retracing. If the dominance of technology today is only the culmination of a long history of the mistaken "double grounding" of ontology onto a theological or primary force, then a "double reading" that allows us to see how that process takes place and to contextualize it within that history might provide some purchase on its power or some insight into its deceptive qualities. Finally, there is Heidegger's suggestion that we need to rethink our standards of valuation, particularly those of how we judge the importance of meditative thought or philosophy proper. If the world is dominated by the "will to will" or a calculative reasoning that focuses on only immediate results, and this is at least partially what allows something like cybernetics to take over the role normally given philosophy, then we must, in some sense, start to value that which is not at least immediately effective, or not effective by the same standards dominant in contemporary culture. As one might expect, philosophy itself becomes key vector in this process, which Heidegger tells us "is not a kind of knowledge which one could acquire directly, like vocational and technical expertise, and which, like economic and professional knowledge in general, one could apply directly and evaluate according to its usefulness in each case" (*Introduction* 9). Indeed, as Heidegger goes on to argue, whenever it appears that philosophy has found "a direct resonance in its [own] time," that is our signal that it is not really philosophy, or it is a philosophy being misused, and those who makes such claims for it are making the mistake of thinking philosophy could be evaluated "according to everyday standards that one would otherwise employ to judge the utility of bicycles or the effectiveness of mineral baths" (9, 13). In other words, insofar as the cult of pure functionality is the sign of the absence of "real" self-understanding, philosophy is most apparent to us in the ways in which it does not "work" or "function" along these pragmatic or utilitarian lines.

And it is here, in Heidegger's three strategies for responding to contemporary technology, that I think we might detect a large number of correspondences with contemporary thinking about the problems of (today's) modernity. Perhaps most obviously, we might find some fairly clear parallels between Heidegger's counterposing of the hypermediated and decadent present with an original or "rooted" centering of human life and society with a wide variety of conservative calls for a "return" to a more grounded or simpler way of life. Indeed, the very symptoms that mark the degradation of the human spirit under the reign of technology for Heidegger in "The Age of the World Picture"—"the flight of the gods" and secularization of the world, the reduction of all community bounds to those of "mere culture," the triumph of hollow consumerism and the abandonment of tradition—reads much like the laundry list of complaints issues by a variety of what we might call contemporary "retro-fundamentalisms," from religious radicalisms and ethnic chauvinisms to the (comparably benign?) calls for a return to "simpler" styles of life and value systems. While it may go too far to suggest that a large spectrum of contemporary conservative ideologues are kissing cousins of Heidegger, there is more than a mere family resemblance between his more elegiac or nostalgic mode and recent "back to X" movements in the present political scene.

At the ostensible other end of the continuum, we might find a significant legacy of Heidegger's "double reading" strategy, his zeroing in on the transitory nature of what has historically served as an ostensibly transhistorical arbiter of truth and reality, in left-oriented social and political theory. Here, of course, there is a particularly direct connection between Heidegger's thinking and leftist cultural that theory drew some of its most powerful concepts from continental philosophy of the last few decades. Indeed, what we might reasonably consider the two greatest content providers for this domain, Jacques Derrida and Michel Foucault, seem to have both found appreciable influence in Heidegger's "ontological history." As numerous commentators, including Derrida himself, have pointed out, Derrida's immensely influential concept/practice of deconstruction is something of a belated continuation of Heidegger's planned *destruktion* or "the task of destroying the history of ontology."[1] While Foucault's genealogical work on the history of knowledge owed much clearer debts to Nietzsche, he would state in his last interview that Heidegger had always been for him "the most essential philosopher" and that his "entire philosophical development was determined" by his reading of Heidegger's works ("Return" 250). And indeed, Foucault's investigation into what he himself has called a "historical ontology of ourselves" would follow a similar path—from the history of European thought back to Greek antiquity—and methodology as Heidegger's, even if its upshot for present day readers, one Foucault described as showing people that they are "much freer than they feel," was almost entirely the opposite of the one offered by Heidegger in his writings on the present state of ontology ("What" 20).

More generally, it is certainly also worth mentioning the degree to which Heidegger's prizing of "authentic" [*eigentlich*] modes of self-relation and action has cast a long shadow over contemporary thinking about related modes of "self-actualization" or "resistance" to dominant modes of social subjectivity or behavior. While the term "authenticity" itself may be long out of fashion in much left-oriented academic discourse, the qualities associated with the concept have largely returned under new names. As Jeffrey T. Nealon argues, for instance, what we might call the "agency question" in contemporary cultural theory is really a disguised "question about 'authenticity'"; when the question of agency is raised in critiquing a theory of power as either too totalizing, or too naively volunteerist, "agency is a code word for a subject performing an action that matters, something that changes one's own life or the lives of others . . . doing something freely, subversively, not as a mere effect programmed or sanctioned by a constraining social norms" (*Foucault* 102). In this context and several others—the turn to affect in social theory discussed previously in this book would also be apposite here, for instance—what Adorno once called the "jargon of authenticity" in Heidegger's work remains, in many ways, the *lingua franca* of theory in the social sciences and humanities, particularly any that take up questions of power and politics explicitly.

Finally, we might see a legacy of Heidegger's emphasis on the "nonoperational" character of philosophic thought—and in a sense also gather together both the left- and right-leaning pseudo-Heideggeranisms sketched above—in the context of the "death of theory" debate we had occasion to reference in this last chapter. Insofar as this debate has revolved around presumptions that theory has become (or always was?) "useless," and/or that its central tenets have become misused by the kind of entities and institutions it was originally designed to critique, I take it to be one very much bound up with Heidegger's earlier concerns about the misuse of philosophy and the need for theory to be meditative and not immediately impactful discourse—the counterpoint to a social realm that too quickly fetishizes the immediately useful and effective over everything else.

At the same time, however, these more recent reflections seem to ask somewhat different questions about the viability of "impractical" theory as a counterpoint to contemporary cultural paradigms. Consider, for instance, *Critical Inquiry* editor W. J. T. Mitchell's response to the *New York Times* article on the journal's 2003 symposium that in many ways set the ball rolling for the "death of theory" discourse. In what Mitchell himself calls "an embarrassingly earnest letter" he sent to the *Times* in reply to the article, Mitchell tells the *Times* editors that their original article title ("The New Theory Is That Theory Doesn't Matter") is not so much inaccurate as it is lacking a crucial qualifier: "immediately." While the *Times* coverage accurately described the symposium participants' negative response to the question of whether cultural theory could immediately intervene in the Iraq War taking place at the time, it failed to acknowledge that cultural theories "take time to percolate down to practical application" (328). Offering a particularly appropriate example, given the context, Mitchell goes on to suggest that "those who think theory doesn't matter should note that the present war in Iraq is the long-term consequence of political theories (hatched at the University of Chicago, amongst others) that are now heavily represented by key intellectuals in the Bush administration."

What is striking to me about Mitchell's response here is how it hews toward Heidegger's consistent emphasis on critical thought as a preparatory and largely anticipatory activity, one whose practical effects, if any, are largely only apparent long after the fact, as well as how it also introduces, at least by proxy, one of the other major planks in the "theory is dead" platform: that left-oriented cultural theory is in decline precisely because of the availability of its central tools for appropriation by actors of almost any type, that theory has indeed become itself as operational or pragmatic as the forces it was designed to counterpoint and slow down. In this sense, then, critical thinking seems to be itself very much a contemporary technic, or part of the "essence of technology" today, just as much as the aleatory consumerism and proliferation of spectacle that Heidegger railed against in his own time. In this sense, turning the crank on the history of ontology another notch, we might say that even Heidegger's critique of the essence of technology and its legacies today look more like a *symptom* of, rather than a strategic *response to*, the contemporary regime of technics, or, to put it a little more pointedly, Heidegger's *solutions* to the calculative regime of sociotechnical systems seem to be the *problems* or challenges that are posed by its current incarnation.

In any case, it is with that ironic situation in mind that I proceed with my own history of sorts in the remainder of this chapter. Although I will be surveying largely the same terrain that Heidegger covers in defining his "epochs" of ontological history, I want to shift ground somewhat in deciding precisely what convergence of forces should take priority. Despite the fact that Heidegger's work in and around the history of metaphysics is in one way all about mediation—the ways in which our access to the original pre-Platonic question of Being has been undermined by the mediation of history, the ways in which our present day experience and self-understanding suffers from a certain overmediation of cultural and technological forces—it is, in another way, almost entirely blind to questions of media; in other words, while Heidegger may attend to a broad contour of the evolution of technology and its impact on cultural systems of (self)representation (from his famous examples of hand tools in *Being and Time* to his writing about early computation technologies and cybernetics), there is very little discussion of what today we might take to be a major vector of contemporary technology: the ways in which it has become intertwined with a variety of *representational media*. Thus my own intervention or refraction here is by way of studying what might be more appropriately considered a "history of techniques," a tracing of the convergences between media and technics or technology, as opposed to the combined history of ontology and technics favored by Heidegger. Such a shift in focus, it seems to me, has much to teach us about rethinking our responses to contemporary intersection of media and technology and its cultural impact. More precisely, if an ontological history results in telling us something about the progression or regression of epistemic frameworks and its effects on our contemporary modes of self-understanding and ontological possibility—Hegel's tracing of the coming-to-consciousness of human freedom, Heidegger's exposure of the "nihilistic freedom" of mid-twentieth-century culture, or Foucault's mission to show people they are "freer" than they feel—then the alternative offered here, while still touching on how what me might call "human freedom" in a generic sense is managed in contemporary cultural formations, will have a more sustained focused on the particular techniques dominant today, how they function and might be made to function otherwise. Considered from a different angle, this sequence might also be viewed as substituting Heidegger's narrative of the historical decline of philosophy and critical thought as an organizing principle for culture with one that indexes the increasing centrality of persuasion, charting not the decline in the force of philosophical reason within culture, but the return of rhetorical forces that were originally crowded out in the formalization of Western philosophical thought.

More programmatically, in place of Heidegger's five epochs, this chapter's periodizing schema will focus on three key historical shifts in the arrangement of media and technics.

From the Measurement of Transhistorical Qualities to the Measurement of Universal Quantities

As numerous historians of science have noted, the emergence in mid-thirteenth-century Europe of diverse new systems and tools for the quantification of phenomena of various types inaugurated a fairly radical break with the legacies of Platonic and Aristotelian metaphysics and early Greek science. According to the classicist F. M. Cornford, the so-called Greek miracle, the emergence of Greek science and of "rational" thought and the retreat of a mythological worldview, was motivated by a new desire to formulate "an intelligible representation or account (*logos*) of the world, rather than the laws of the sequence of causes and effects in time—a *logos* to take the place of *mythoi*" (141). The resultant focus on transhistorical substances or qualities would long dominate science and intellectual thought in Western Europe for many centuries, until the proliferation of techniques for perceiving and representing phenomena as variable quantities that found its start around 1250 BCE. The historian Alfred W. Crosby, appropriating a term that goes back until at least the mid-fifteenth century, has referred to this era as one marked by an urge toward "pantometry," the systematic measurement of all things. Key to this sea change would be the creation of a variety of instruments (from accounting tables to public clocks) and representational techniques (the integration of geometric perspective into illustration and the refinement of cartographic projection) that trouble the barriers between phenomena that represent information and machines that perform technical processes.

From the Determination of Quantitative Measurements in the Present to the Calculation of Probable Relationships in the Future

During the start of the seventeenth century, roughly coinciding with the formalization of the mathematical calculus by Newton and Liebniz, we see the emergence of a large number of new systems and practices that move beyond the measurement and relation of discrete quantities to determining, on the one hand, more flexible relationships of correspondence, compossibility, or adequality, and, on the other, to conceiving of the relationships between phenomena as a series of processes unfolding in time, sequences whose probable results could be calculated if enough of its variables could be determined in advance. The statistical information and data relationships of pantometric techniques here become operationalized in a series of social processes and management styles for political economy that might be taken to have peaked in the early to mid-twentieth century.

From the Calculation of Probable Relationships to the Creation of Algorithmic Processes for Managing Associations

This contemporary moment, one largely midwifed by the cybernetics movement in the mid-twentieth century, is marked by a shift toward the integration of a variety of flexible systems in which the pursuit of specific goals is secondary to maintaining a systematic equilibrium that permits a spectrum of "acceptable" results to take place. This particular shift, I will argue, is at the heart of a number of seemingly paradoxical changes in contemporary culture and political economy that addressed in the previous chapter—from the "standardization" of niche marketing, to the loosening of prohibitive strictures as a control mechanism within social power.

   In what follows, I address the first of the sequences in reference to its importance as a genetive context for thinking about media, technics, and culture in Western intellectual thought, and the second because I believe it to be the regime that largely remains the focus of contemporary criticism of technology. However, as a reader might expect, my primary concern will be the final sequence, one that I take to be the dominant cultural paradigm for the contemporary convergence of media and technics today.

PANTOMETRICS

Although it may appear counterintuitive, the convergence of media and technics in the period with the greatest historical distance from the present may also be the one that is most familiar to modern minds. While the methods for standardizing measurements and their representation that was the driving force of immense industrial, scientific, and artistic achievements at this time—from mechanical clocks, to accounting tables, to the introduction of geometric perspective into visual representation—may have been spectacular and novel applications of human-made frameworks to natural phenomenon, they have become so ubiquitous and proletarianized in the present as to seem entirely natural, if not invisible. Similarly, the conceptual process that helped transform the prevailing *mentalité* of Western society beginning in thirteenth century and that was embodied in such devices is now largely second nature. As Crosby describes it, the spread of universal measurement required mastering only a simple three-step process: "reduce what you are trying to think about to the minimum required by its definition; visualize it on paper, or at least in your mind, be it the fluctuation of wool prices at the Champagne fairs or the course of Mars through the heavens, and divide it, either in fact or in imagination, into equal quanta" (228). For these reasons, perhaps the only way to mark the paradigmatic nature of such a shift—and for our purposes here, to begin trailing a recurring process that we might map on to successive regimes of technics and media—is to consider its distance from the opposing emphasis on *qualification* that was inherited from Greek intellectual culture.[2]

Insofar as the predominant vector of scientific and industrial representation in the late Middle Ages and early modernity focused on the presumed universality of measurement schemes, this was a radical break with the very concept of "universality" inherited from Greek antiquity, as well as from the ways the Greek themselves measured the relative importance of disciplinary domains of knowledge. While using the term "qualification" to describe Greek approaches to metrics has the advantage of contrasting nicely with the emphasis on "quantification" that came to later dominate scientific thought, it is perhaps more enlightening to consider how Greek intellectual thought's investment in determining the essential qualities of phenomena led to a dominant concern with absolutes and transhistorical essences. Plato's concentration on determining the necessary qualities of both physical objects and mental concepts is, of course, a key vector of his dialectical method, one often linked with his equally consistent argument that "the truest of all kinds of knowledge" was produced only by disciplines "concerned with being and with what is really and forever in every way eternally and self-same" (*Philebus* 58a). Thus in the *Philebus*, amongst other dialogues, arts and sciences that study variable phenomena or historical mutability itself are relegated to an inferior position, with rhetoric—that discipline that, on Plato's reading, pretends to universal application via a counterfeiting of philosophical methods—singled out as a particular danger.

Against this backdrop, the sea change initiated by quantitative reasoning and the popularization of pantometric media (geometrically based illustration and cartography, musical notation) and devices (calculating boards and accounting tables, public clocks) is perhaps most striking in how it disturbed this rigid discrimination of *episteme* from *techne* that was inherited from Greek culture; these phenomena seemed to simultaneously perform or in some ways combine functions that used to be exclusive to representational media—particularly those that would record or present established knowledge or logical concepts and relations—with that of physical technologies that produce actions in material reality. In this sense, period technologies such as calculating tables were both means of representation and tools for the actual performance of measuring and relating different quantities. For their part, more recognizably static signifying media such as uniform musical notation and reliable cartographic projections were emblematic of an increasingly "executable" form of representational media, one in which its representational capacity was only useful insofar as it was operational for its users. Both types would perform the dual purpose of being both representational media and technologies for establishing relationships between individuals and items, mediating economic exchanges, and establishing cultural norms.

We might take the installation of large mechanical clocks as a synecdoche for the pantometric conflation of media and technics in at least three significant ways. First, the standardization of shared time that took place via the installation of public clocks demonstrates a clear break with the purely representational or static mimetic capacities of previous media. Importantly, such a break would come through a modeling of dynamic organic processes within a mechanical realm. The presumption of a constancy of "time" that would inspire the making of something like a mechanical clock is, of course, based on a particular appropriation of natural processes (the division of time by the orbit of the earth as well as the "body clock" of circadian rhythm). The "automaticity" of the mechanical clock designed around its approximation of biological function would be both an early avatar of later modeling of living systems within technical realms as well as an inspiration for what E. J. Dijksterhuis would call the "mechanization of the world picture" in science and philosophy of the sixteenth and seventeenth centuries, a way of representing the world or various natural and social systems within it as a discrete entity, but one composed of moving parts changing over time (both Descartes and Leibniz, for example, would come to use mechanical clocks as key symbols for life and consciousness in their metaphysical writings).

Second, public clocks of this time are also one of our best examples of the ways in which pantometric devices created a *socialization* and a *secularization* of forces and concepts that were previously "held" privately or under the aegis of religious rituals and authorities. Time was of course already "shared" socially before the installation of public clocks via a variety of mechanisms used to broadcast the hours kept by private timepieces, but public clocks both regularized and for the most part removed the need for constant human intervention into the process of keeping public time.[3] Thus the introduction of public clocks in the Middle Ages proletarianized a function that used to be largely regulated by religious authorities; as the historian Jacques Le Goff writes, the introduction of public clocks led to "the great revolution of the communal movement in the time domain," as merchants and artisans "began replacing this Church time with a more accurately measured time useful for profane and secular tasks, clock time" (35). More generally, this phenomenon led to what philosophers of science Isabelle Stengers and Didier Gille call "the autonomization of social time," a sociotechnical event that will play a key role in prompting further investigations into the best ways to establish "universal" systems of chronological and spatial projection that can be shared inside and across communities of various types, a "second nature" of temporality that could take the place of less reliable systems of measuring astronomical time with sundials and other instruments whose accuracy was dependent on the skill of their individual users (179).

Finally, the popularity and ubiquity of socialized time embodied in public clocks provides perhaps our best demonstration of the ways in which pantometrics shifted the economies of quality and quantity as conceived in Greek antiquity, as well as helped augur the return of the *metron* techniques of early rhetorical and sophistic thought opposed by Platonic metaphysics. The abstraction made possible by quantification became prized for its functional, rather than epistemic accuracy; while systems for producing and relating quantities of various types were obviously "artificial" in the sense that they did not represent any "inherent" or "essential" property of the phenomena being measured, this is what formed, as Henri Bergson would later suggest "the quality of quantity" that would allow one to "form the idea of quantity without quality," or a measuring system based solely on effectivity rather than epistemology (*Time* 123). While in Heidegger's schema we might retrospectively see a shift in epistemic and pragmatic prioritization as a precursor to the "essence of technology" that takes root even earlier than Heidegger may have considered (the shared time of public clocks being an early version of the rejection of "individual time" that Heidegger attributes to radio broadcasts?), it is perhaps best to consider the withdrawal of concern for ontological properties in favor of pure function as an early step in the sequence through which ontology will increasingly be itself *put to work*, a process through which human conceptual power and the prevailing *mentalitié* of a community is actively integrated within cultural processes.

PROGNOMETRICS

Near the start of the seventeenth century, the pantometric techniques and large variety of devices and systems based on *quantification* began to give way to an entirely new system of techniques, technologies, and models of mental operation based instead on *calculation*. Indeed, the end of Crosby's periodizing scheme corresponds roughly to the formalization of calculation proper in mathematics through the co-invention of the infinitesimal calculus by Newton and Liebniz in mid-seventeenth century. We might take one of the key concepts of the that system, *adequality* or the concept of "proximal equality," as a suturing point for a large range of subsequent cultural changes based on the assumption of "likely" relationships between phenomena or the results of particular processes. The emergence of a number of techniques of this time that we might describe with the portmanteau word "prognometrics" ("prognosticate" or predict + "measure"), would demonstrate a turn away from methods for quantifying phenomena in the present and toward procedures for predicting the probable future states and effects of natural, mechanical, and social systems of various types. The effects of phenomena occupying a state between mediation and technology here and its impact on societal change during modernity might be found, for instance, in the increase in stock-exchange systems and early speculative markets, as well as series of new techniques and technologies, spectacles and sign systems, used to define social roles and modulate the future behavior of individuals that perhaps found their most influential study in Foucault's *Discipline and Punish*.

In this sense, prognometric techniques might be taken as a break with pantometrics, but one that intensified the latter's initial movement away from traditional metaphysics. Key to this process, as with the shift from the Greek emphasis on qualification to the quantification refined over the three previous centuries, the emergence of prognometrics in the seventeenth century would mark a further remove from the consideration of ontology as a "first philosophy" that could account for the eternal nature of things, and toward the strategic leveraging of human sensible and ideational capacities as an operational component of industrial, aesthetic, and political design. Beyond its important role in the history of mathematics proper, the development of the calculus might taken as a conceptual linchpin for an entire host of practices and design strategies that may have found their roots in the rarified world of metaphysics, but would soon spread into and beyond art and industry to become a quilting point of social life in the subsequent century.

As the French mathematician Rene Thom suggests, the controversies between the followers of the physics of Descartes and Newton that were reaching their apex at the end of the seventeenth century were ones largely arranged around the question of whether the ontological concerns of traditional scientific theory or the more immediately pragmatic forms of abstraction based on calculation should be given precedence: "Descartes, with this vortices, his hooked atoms and the like, explained everything and calculated nothing; Newton, with the inverse square law of gravitation, calculated everything and explained nothing" (5). Insofar as pantometric methods and technologies overcame the Greek obsession with transhistorical absolutes and the unchanging properties of substances through time, prognometrics would accelerate this process. Most notably, the quantifiable regularities of both the natural world and the operations of social systems that became the common coin of sixteenth-century science would be turned into mathematical variables that would not only represent the common behavior of these systems, but also predict their likely future behavior under specific conditions.

As with the switch from the commonplaces of Greek metaphysic and *mentalité* to the natural philosophy and conceptual tools of pantometrics, this shift might also be discovered at least provisionally in changes in formal scientific methods of the time, and then subsequently traced through the innovations in art and aesthetics as well as the material technologies of the same moment. If the Aristotelian legacy within medieval science could be seen most strongly in a disaffection for observational methods and physical experimentation, and the pantometric era can be defined around the direct measurement and manipulation of imaginary units of quanta and the use of specialized tools for manipulating such units, prognometric methods and techniques would, on the one hand, demonstrate both an internal accounting for human variability in the use of measuring technologies (and eventually its full-scale automatization), and, on the other, display an increase in the forecasting abilities of such measurements to not only record past and present behavior but provide oversight on future planning. The former would largely take place through a great formalization of standards for calculating and extrapolating conclusions from vast amounts of quantitative data—the emergence of statistics as a recognizable discipline—the data itself in many ways only newly enabled by the massive increase in material records made possible by the popularization of printing technologies.

Such a process pushed the pantometric desire for "universal measurement" to another level—now there would be not only universal units of measurement, but also universal standards for how one would aggregate and extrapolate conclusions from such measurements. As many philosophers and historians of science have noted, such mechanisms emphasize a growing awareness of the subjective element of human observation and record collection—an attempt to acknowledge and account for the potential errors on behalf of individuals collecting such data. At the same time, though, corrective protocols appear based on the trust that revised methodologies could ameliorate such problems, pursuing what the historian Zeno G. Swijtink has called "observation without an observing subject" (280). At the same time, however, there was an increased concentration on methods for dealing with outliers and unusual results; the concern for accounting for such irregularities was in many ways a result of the increase in quantitative data available, but it was such increases in the size of data that made mathematical methods for dealing with irregularities a viable strategy. The result is what we might call an increase in the "operationality" of quantitative information of this type. For instance, although vital listings were kept in much earlier times, by the 1600s "bills of mortality" would be made publicly available so that, as the historian Richard H. Shyrock writes, "emerging dangers could be detected and that the genteel would 'know when to leave town'" (225).

The ability to manipulate quantified information would lend a different type of "operationality" with aesthetics and art proper of the time, one that would intensify the tendencies toward the simulation of living systems within inorganic realms as well as the increasing conflation of the domains of technics and media generally that found its formal start in pantometric techniques. This is perhaps most clear in Baroque art; indeed, the key defining traits of Baroque art and architecture—an abundance of fine details, an emphasis on lighting effects, the use of *trompe-l'œil* and other visual illusions simulating dimensionality, movement, and the images approximating multiple generic forms simultaneously—speak to the use of formal (mathematical) and informal (strategic) techniques of calculation far beyond the introduction of perspective into images, concomitantly increasing capacities to approximate movement and "liveliness" into static material works. Heinrich Wöfflin would summarize these tendencies in his early and influential treatment of the Baroque around a displacement of traditional geometric forms by ones that spectators would experience as flickering between different states, or appearing to move despite their solidity:

> Line was abolished; this meant in terms of sculpture that corners were rounded off, so that the boundaries between light and dark, which had formerly been clearly defined, now formed a quivering transition. The contour ceased to be a contiguous line; the eye was no longer to glide down the sides of a figure, as it could on one composed of flat planes ... Just as no need was felt to make the lines of contours continuous, so there was no urge to treat the surface in a simpler way; on the contrary, the clearly defined surfaces of the old style were purposely broken up with 'accidental' effects to give them greater vitality. (35)

Such tendencies might be taken as synecdoches for two ways that calculation became a conceptual operating principle for a wide variety of techniques and endeavors of various types that newly emerged in this period. On the one hand, there is the explicit use of *mathematical calculation* in the design and creation of these architectural features; on the other, there is also an emphasis on *calculating the experiential effects* of the created design on its future viewers, an introduction of the physiological sequence of perception and reaction on the part of the viewer into the design process. As Wölffin writes, because the viewer of the detail-packed and abstract contrasts of Baroque visual works "cannot possibly absorb every single thing in the picture" they are "left with the impression that it has unlimited potentialities, and his imagination is kept constantly in action" (34). If pantometric techniques of exact and "usable" measurement helped bring into existence a new type of art that could more precisely approximate the surface and contours of organic objects, the introduction of more advanced mathematical methods and the general interest in calculating procedural effects into the design of Baroque works would provide a way to stimulate organic movement, and to more precisely presume and dictate the organic process of visual perception and affective response to such works.

Through the movement in Baroque architecture in which decorative elements became more specifically communicative, and through which Baroque artists of various types integrated the presumed sequence of perceptual processes that would be prompted by such materials, we might recognize the distinct ways in which what I have been calling prognometric techniques demonstrate a merging of technics and media that would be a step beyond what was achieved in the pantometric era. This conflation, and the procedural character of Baroque representational media more generally, might also serve as the principle that connects it closely with coeval developments in science and metaphysics of the time.[4] The calculating and tracking of the mental processes of the viewer would find a complement, perhaps most obviously, in the Associationist School of psychology and philosophy; as some historians of science have suggested, concepts of the human mind and perceptual process as described by such thinkers as John Locke, David Hume, and David Hartley are closely linked to more general intellectual concerns with probability and the forecasting of possible effect: both considered "the mind a kind of counting machine that automatically tallied frequencies of past events and scaled degrees of belief in their recurrence accordingly" (Gigerenzer 9). In a more methodological vein, Dalibor Vesely suggests that the abstract visual forms produced by Baroque luminary Giovanni Battista Parinesi presumes an operation of mental abstraction and processing on the part of all viewers that might be compared to that of the "thought experiment" now common in empirical science; much as a thought experiment presumes that a participant can imagine an idealized version of reality, the distinctive qualities of baroque visual art rely on a similar power of abstraction and real-time cogitation by its viewers (257, 443n49). The spread of such techniques into the general social field demonstrated a further movement in the ways ontological conditions of the era became available as strategic resources for unique techniques and technologies of this moment.

FROM PROGNOMETRICS TO PARAMETRICS

I spent a considerable amount of time above tracing the emergence of statistical methods in social governance and the origins of "calculative" economies for a very specific purpose. While the use of statistics in the management of political economy go back at least as far as the early seventeenth century and "machinic" perspectives on social management are largely composed in the late eighteenth century, I take it that it is these types of theories and practices—ones in which quantifiable data and mechanical calculation are the coin of the realm—that are typically marshaled as evidence in discourses that identify contemporary (information) technology and a reliance on quantitative research methods as, if not the outright cause, then at least adroitly symbolic of a society taken to be increasingly inhumane or mechanistic in its logic. In other words, in addition to Heidegger, it seems like a vast majority of critiques of this kind are premised on the notion that we have abandoned some socially ameliorative and humanistic investment in critical judgment in favor of a world system run on brute calculations.[5]

Perhaps a touchstone for this critique is Adorno and Horkheimer's *Dialectic of Enlightenment*, a text that takes the contemporary domination of calculation or programmatic reason as exigence for tracing its beginnings back to Enlightenment thought. For Adorno and Horkheimer, when a philosopher like Kant, the figure who perhaps more than any of his time made the concept of judgment central to philosophical inquiry, ends up "confirming the scientific system as the embodiment of truth," this starts the beginning of the end of judgment as it might be conceived in any ideal or even positive sense; the self-reflective capacities crucial to judgment disappear and "thought sets the seal on its own insignificance," emulating scientific reason and thus becoming "a technical operation, as far removed from reflection on its own objectives as is any other form of labor under the pressure of the system" (66). While tracing this tendency back to the eighteenth century when advances in mathematics and formal logic "offered Enlightenment thinkers a schema for making the world calculable," it might be taken to be approaching its apex in the postwar period, as shown in Adorno and Horkheimer's famous descriptions of the impact of positivist methodologies on social analysis and government planning and that of the "the Culture Industry" in shaping human motivation and attribution of value (4). Today's world, or at least the world of early postwar America, is one in which "chance itself is planned": "For the planners it serves as an alibi, giving the impression that the web of transactions and measure into which life has been transformed still leaves room for spontaneous, immediate relationships between human beings" (117).

However, as I have already had occasion to suggest in the introduction to this book, it might equally be said that the wartime period—such a central symbol to Adorno and Horkheimer in *Dialectic of Enlightenment* as it was, sometimes for very different reasons, for the late work of Heidegger—was also the birthplace of a variety of prototypical techniques that would radically revolutionize not only contemporary technology and media, but also a wide variety social and economic structures. These new techniques would follow the pattern of change and intensification that I have been sketching above, but would alter the logics of calculation or what I have been calling prognometrics in such a fundamental manner that it would be very hard to say that the targets of either Heidegger's or Adorno and Horkheimer's analyses remain dominant in the present moment. I map the occurrence of these *parametric* techniques throughout a variety of domains later in this chapter, but first I will try to unpack the difference between critical conceptions of the "calculative" or "reductionist" impact of technology on culture and what I take to be the more dynamically processual or algorithmic logic of society after cybernetics.

We might find a particularly resilient example text for such an objective in a work composed just a few years after the publication of Adorno and Horkheimer's masterpiece. Readers of Alan Turing's 1950 essay "Computing Machinery and Human Intelligence" can be forgiven for presuming that they were about to read an argument by one of the century's preeminent logicians about how the second subject of the title ("human intelligence") might be realized in the realm of the first ("computing machinery"). Though computer science research had certainly yet to produce a machine that could rival human intelligence at the time of Turing's writing, human expectation—if not technology—had reached science fiction proportions. However, in the essay Turing immediately brackets the question "can machines think?"—a query he immediately indicts as "absurd," possessing a "dangerous attitude," entailing an infinite regress of definitional obligations, and "too meaningless to deserve discussion"—and attempts to convert the thrust behind this query from an abstract concern into a performance containing multiple practices of human-machine interaction, mediation, and response (42–43). In place of a traditionally epistemological concern with the nature of "thinking" Turing suggests a retrofitting of the popular party pastime, the "imitation game." The game is normally played with a man, a woman, and an "interrogator" of either gender. The interrogator asks the other two participants a series of questions, their responses are mediated through a third party or preferably transcribed or typed, and then the interrogator attempts to determine based on these responses alone the gender of each participant. Turing hypothesizes that the substitution of a "thinking machine" for one of the respondents in this game would create a more productive environment for engaging the significance behind the original question of "can machines think?"

Rather than attempting to "prove" the identity of human and machinic thought by recourse to formal logical or definitional mechanisms, or his experience as a leading mathematician and logician of the era, Turing's retrofitted imitation game relocates this process to an interactive experiment that requires the reader's (virtual) participation. As a game, Turing's test constitutes a series of algorithmically complex rhetorical practices of interaction and persuasion that hinge more on human response to machinery than any philosophical inquiry into the possibility of a "thinking" machine: "The original question, "can machines think?" I believe to be too meaningless to deserve discussion. Nevertheless I believe that at the end of the century the use of words and general educated opinion will have altered so much that one will be able to speak of machines thinking without expecting to be contradicted" (442). The original query ("can machines think") immediately hails questions of agency and cognition traditionally proper to philosophical reflection, an inquiry Turing pejoratively describes as one "best answered by a Gallup poll" (433). His rendering of it into a game depends no less on human response and no less on the quantification appropriate to polling—"I believe that in about fifty years time . . . an average interrogator will not have more than 70 per cent chance of making the right identification after five minutes of questioning"—but these effects emerge out of the material practices between human and machine when put into circuit with one another and assay not the identity between the silicon and flesh participants but the negotiation of their alterity (442).

Turing's essay was the virtual rehearsal for an experiment in the clinical sense of the word and entailed all that goes with such a process, such as reproducibility and modification. As such, the original Turing Test comprises only the first iteration of a process that has been relentlessly duplicated and modified during the past half-century in actual attempts to assay artificial intelligence programs. In addition to this physical repetition, the Turing Test is often reiterated textually as a point of departure for gauging rapid advancements in automated computing technologies and the related rise of informatics in the life sciences. In the latter endeavor, the Turing Test, in both its technological and cultural import, has been popularly thematized as a pivotal event in not only the modeling of machines on the basis of human behavior, but also the reflexive erasure of embodiment in theories of human cognition and subjectivity: if a machine can think, or at least fool an interrogator in the very pragmatic performance of the Turing Test, then surely, many critical interpretations suggest, the difference between the two can be reduced to so much vitalist hand-waving, and intelligence and consciousness have nothing to do with the very human quality of embodiment and the very humanist quality of self-sovereignty. Hayles, for instance, positions the Turing Test as a ground zero for cyberneticists' supposed dismissal of embodiment from considerations of consciousness: "If you cannot tell the intelligent machine from the intelligent human, your failure proves, Turing argued, that machines can think. Here at the inaugural moment of the computer age, the erasure of embodiment is performed so that 'intelligence' becomes a property of the forceful manipulation of symbols rather than enaction in the human lifeworld" (*Posthuman* xi). For Hayles, the dislocation of the question "can machines think?" from a definitional query to the operative process of his revised imitation game does not so much shift the grounds of his engagement, but compels the reader to accept deceptive assumptions: "Think of the Turing Test as a magic trick. Like all good magic tricks, the test relies on getting you to accept at a certain stage assumptions that will determine how you interpret what you see later." Similarly, in a slightly more nuanced reading of the role of embodiment in Turing's test, Elizabeth A. Wilson argues that the "attentive constraint and management of corporeal effects" in the test produces the "projection of a certain fantasy of embodiment—specifically, the possibility of noncognitive body, or (in what amounts to the same thing), the possibility of a cognition unencumbered by the body" (111, 120).

At first blush, all of this—the association of Turing's test with the perceived disembodiment of human intelligence or human subjectivity, its positioning as a key moment in the history of science and technology fields that would come under fire by humanists and social scientists in the twentieth century as reductionist and politically naïve—makes a lot of sense. Turing's does seem to be suggesting that something like "thought" or "intelligence" can be captured in written form (the pieces of paper passed back and forth from the "test subject" and its hidden interlocutor), he does go to great pains to restrict any test of a computing machine that would require it to have a (visible) body, and, as his appropriation of the "imitation game" based on guessing gender implies, the true test of whether an observer will presume they are interacting with another person is very much reliant on how well the machine is able to exhibit what will be accepted as (stereo) typical behavior for a human.

On a second read, however, one that pays closer attention to the specific parameters of Turing's Test and its processual design, there is also good reason to be surprised that cultural critics, particularly those well versed in contemporary theories of the performativity of identity categories such as gender, would find Turing's experiment so grating. After all, it would appear that it is the exact notion that our perception of human identity is premised on a series of repeated performances of social typologies and behaviors—from Nietzsche's famous suggestion that "there is no 'being' behind the deed . . . 'the doer' is invented as an afterthought" (*Genealogy* 26), to the work on iterability and the "practices" of identity in Derrida, Foucault, and Judith Butler's groundbreaking writings—is pretty much the conclusion that put the "post" in postmodern theory.

In other words, while there might be solid ground for such critiques if Turing were making a strong argument for an explicit and comprehensive definition of "actually existing human intelligence" in his essay, but given that Turing's interest is in testing how well a machine might approximate or reproduce such culturally conditioned factors and commonplaces in order to fool an interlocutor into accepting an incorrect identity, it seems that

Turing's test is more itself a conscious acknowledgement of and program-matic response to postmodern notions of perfomativity. Indeed, we might even go further and state that Turing's test was not so much the start of reductionist, "disembodying," and conservative notions of human subjec-tivity that might insist on rigid conceptions of human identity and appro-priate human behavior, but rather the beginning of a process in which the wide range of possibilities in these realms are integrated into systems of communication and interaction within and between human and techno-logical realms.

For these reasons I perform my own restaging of the Turing Test here in an attempt to remobilize the dynamics of Turing's engagement some-what differently than is common in analyses of its cultural legacy; specifi-cally I am interested in considering these vectors not as a *epistemological* endeavor but as a *rhetorical* ecology, if we can take rhetoric, as I suggested earlier in this text, to name not just economies of persuasion in general but the a variety of techniques based on flexible formulas that eschew logics of identity and difference in favor of "as if" and "if then" directives and procedures. Even more precisely, we might say that Turing's schema here is not a philosophical argument that hypostatizes what "counts" as cog-nition, denigrates the importance of physiological bodies in this process, or anticipates the appearance of human intelligence in machine domains. Instead it is a strategic description of particular forces in interaction that anticipates much of contemporary culture, one in which the distribution of agency and subjectivity into organic and machinic realms is becoming increasingly ubiquitous and the dominant logic of virtually every important system of interaction—political, economic, cultural—seems to be follow-ing much more the procedural operations of the Turing Test rather than its presumed reduction of human intelligence to operating symbols. In other words, I want to invert the relationship above where rhetoric is taken as a secondary operation of argumentative arrangement that eases the accep-tance of an epistemological conclusion, and instead position as primary the rhetorical arrangement of capacities to affect and to be affected. More specifically, I want to trace a movement of technological evolution in the realms of mediation and automation, one in which the prognometric focus on the accurate predicting and provoking of a spectrum of possible results becomes formalized into an entirely new system of processes. These pro-cesses, I take it, are the more relevant, one might even say more urgent, legacy of the Turing Test today than its common associations with "reduc-tionist" views of (disembodied) human intelligence and outsize claims for the powers of computing technology. More importantly, however, they might also be used to gain some purchases on understanding the ways that a wide variety of social processes function in the present.

We might coordinate the import of Turing's test in relation to con-temporary arrangements of technology, media, and persuasion around three central points. The first is what we might call a *new ontological framework of interaction* anticipated by Turing. As Lev Manovich has argued, insofar as we can claim that there is such a thing as an "ontology of the world according to a computer," a unique logic for identifying entities and their relationships, it is one composed of two parts: data structures and algorithms (*Language* 223). Although there is a variety of ways particular computer programs may couple or associate these two structures, in rough parlance, any entity identified by a computer—whether it be the kind of explicit quantitative units we typically associated with data or any object whatsoever, including such phenomenon as changes in objects or processes—is considered a data structure. Any routine or task involving data structures is encoded in an algorithmic format, a series of commands that make associations between data structures and execute actions in response to rules-based directions for particular scenarios. Although one might see reductionism in the data structure component of this system—insofar as the representation of something by some other symbol is always a reduction of the "totality" of the original in some way—the potential complexity of the algorithmic component is perhaps the more notable or unique component of computing in the present moment. The creation of a variety of subroutines and nested commands within computer algorithms, as well as the ability to arrange commands around "if then" scenarios or threshold events, marshals a kind of complexity that is in many ways compensational to the simplicity of the data structure as a mechanism. In other words, although our popular image of the kind of activity that takes place in contemporary computing is likely one of a variety of phenomenon being reduced to a series of symbols (such as those inexorable "ones" and "zeroes" that became identified with digital code), the really striking progression of computing technologies has been the increase of algorithmic complexity in the pursuit of making ever more flexible responses to tiny changes in data structures and their relationships.

Perhaps a simpler way of putting this would be state that while our understanding of the cultural impact of computing technologies, or perhaps more pointedly, our fears about its potentially dehumanizing or rigidly mechanical logic or reduction and quantification seeping into social life as a whole, have tended to focus on the "dematerializing" or decontextualizing principles of data structures, the actual social impact of computers and information technology as we have come to experience it over the past several decades has instead been parallel to the algorithmic component of computational processes. It is not the ostensibly binary logic of digital coding and the reductive symbolization of data structures that has become mirrored in contemporary culture, but rather the increasingly flexibly, niche-oriented, and generically goal-oriented logic of the algorithm. Consider, for instance, some of the most famous and influential algorithms of the present: Google's PageRank and AdWords/AdSense systems. The results of a Google search, based on Google's PageRank algorithm, are determined by a wide variety of relationships. These include what we might call "generic" elements such as (1) a complex parsing of the relevance of a online site to a particular search request, (2) that site's traffic and the amount of links to it by other sites, and (3) the evaluation of the "quality" of the site based on a number of factors. However, the algorithm also takes into account a variety of factor specific to data available about the user; although such information as the geographical location as identified by network address have long been part of such calculations, from 2005 onward the search algorithm also started to take into account the previous search history of the user and other available information about the user based on their activity using Google and (more recently) other sites and services, including known relationships between the user and other individuals (as derived from their connections on the Google+ social network). When used in application with Google's AdWords and AdSense algorithms, the primary revenue source of the company, advertising is similarly directed toward the user based on such specific and detailed factors.

What is perhaps most notable about this progression in algorithmic complexity is how it demonstrates a shift toward strategies based on the achievements of *more general* goals by appealing to *more specific* audiences. For example, in Google AdWords and AdSense, the purchases of *any product or service*, rather than *a particular product or service*, is made possible by the complexity of the algorithm used to sort and source marketing media and the ability to tailor such media based on the ever more specific preference of ever smaller numbers of people.

Turing did groundbreaking working in the theory of mathematical algorithms across a wide range of his career, but perhaps the fairly simple modeling of such a sequence in "Computing Machinery" is the best example of this kind of algorithmic process, particularly in how this structure in computing proper has bled into a variety of cultural domains. Turing's model of computing—in which the contribution of its interlocutor, regardless of whatever it may be—is assessed and triggers a response tailored to move the process closer to a desired general objective may not be an adequate model of "human intelligence" but it is one that looks increasingly like the dominant logic of a wide variety of economic, political, and otherwise cultural systems in which the efficient ability to produce flexible responses to increasingly diverse and wide-ranging amount of variables or real individuals is very much, as we shall see below, the ruling logic of the day.

Perhaps this distinction might be clearer if considered in relation to our second point—the Turing Test's restriction of representation and its *replacement of a logic dominated by direct representation or identification with one dominated by pure process.* As we saw above, Turing's dislocation of the question of human intelligence from definition to process relies not so much on any belief in stable classifications of human perception or conception as on definitive lapses in these categories: an epistemological "blindspot" created by the inability of human thought to definitively conceive of "itself," and a practical disruption of human vision, the conceit in Turing's game (as well as in the imitation game he is adapting) wherein a participant is not allowed to "see" her interlocutor. The inability of a participant to see the subject being investigated was of course crucial to Turing's test; this elision both created the parameters for what capacities Turing was attempting to survey—"We do not wish to penalise the machine for its inability to shine in beauty competitions, nor to penalise a man for losing in a race against an aeroplane"—and served as a delimiting function on how the experiment would be staged both imaginatively and materially (435). Although feminist critics of Turing's work have pointed out that such a formulation risks eclipsing embodiment from a working definition of intelligence, it is also the component of the test that makes it transparently an exercise in determining the processes through which something like intelligence might be mapped rather than an attempt to generate a definitive conception of it in some kind of static form.

Turing's decision to model intelligence, or more accurately, what would be perceived as intelligence, by way of an (algorithmic) process is very much against the grain of not only traditional representation strategies that might encapsulate this category as well as a very long history of how "intellectual" knowledge itself, specialized or meta-level knowledge "about" something, is conceptualized and expressed. Namely, Turing's focus on the restriction of vision and representation is very much inverse to centuries-old Western tradition reliant on foregrounding some variation on (visual) perception/representation. This is true from ancient Greek philosophy, during which the capacity to have special insight into phenomenon became known as *theoria* ("looking at" or "gazing at"), to what Martin Jay calls the "ocularcentric" tradition of Western intellectual thought, one in which the correct apprehension of reality is equated with "seeing clearly" and being able to represent or diagram in some fashion the relationships between often invisible or obscure substances and forces.[6] This tendency continues, somewhat paradoxically, even in what Jay dubs the "antiocularcentric" tendencies of critical or anti-foundational approaches to metaphysics and philosophy; from Marx and Engels' famous reference to ideology working like a *camera obscura* that makes "men and their circumstances appear upside down" to more recent metaphors of blindness (Paul de Man), parallax (Slajov Zizek), and the more general interest in the (sublime) "unrepresentational" qualities of diverse areas of aesthetics or experience (Lyotard, Levinas) (14).[7]

This is not to say, of course, that "Computing Machinery" does not "represent" anything, or uses some radically strange form of communication or expression, but what it presents is not so much a description of qualities or figurations but a series of formulas and techniques. In this sense, then, Turing's essay, particularly given the subject matter under review within it—the relationship between computing machinery and human perception and cogitation—is a particularly salient anticipation of new economies of media and *mentalité* that would become dominant in the era of ubiquitous information technology; perhaps more importantly, it might also be taken as an avatar of contemporary systems that rely far less on the management of particular representational schemas and the control of what "counts" as accurate representation or veridical truth and serve to *exclude others*, and more on the effective management of a variety of processes of political-economic interaction that attempt to *include as many* divergent beliefs, values, and modes of decision making as operationally possible.

Third and perhaps, at this point, most obviously, Turing's test can be as marking the new regime of two interrelated processes that we have been tracking throughout this essay: *the conflation of technics and media* and the *modeling of natural systems in inorganic realms*. While Turing's essay is often read as an early document in theories of artificial intelligence (an artificial agent that might make goal- and self-oriented decisions) as opposed to artificial life (the modeling of life systems in artificial environments), Turing's anticipation of human–machinic interaction in "Computing Machinery" best demonstrates what would appear to be a third category, what we might call for lack of a better term the *inorganic ecologies* or *artificial liveliness* that were central subjects of much cybernetics research. Here the endeavor is not so much to the formal pursuit of simulating something that might fit an ideal definition of "intelligence" or "life," but rather the design of object, environments, and spaces of interactivity that mimic particular elements of life systems, such as the maintenance of equilibrium and structural adaptation.

PARAMETRIC CULTURES

We might more formally define the parametric as a conceptual paradigm distributed across a variety of contemporary technological and cultural domains—the spread of the new arrangements between *techne* and *logos* or technics and media anticipated by Turing's test beyond the realm of information technology proper—around four core principles; each of these might be read in contrast to both the earlier regimes of techniques described in this chapter as well as more general critiques of the ostensibly dehumanizing or disembodying effects of digital culture and information technologies (which, as I suggest above, may more appropriately describe the prognometric or "calculative" techniques now being eclipsed). First, powers and purposes that used to be attributed to process of direct representation, whether of an indexical or ideational nature, are transferred to more dynamic processes that work to maintain relationships between a variety of elements. Second, whatever we might call social conditioning or "social power" works primarily through logics of inclusion rather than exclusion, an attempt to integrate and draw value from as many heterogeneous identities or behaviors as possible. Third, in control systems of any type (technological or social), the pursuit of specific goals is secondary to maintaining a systematic equilibrium wherein a spectrum of "acceptable" results will likely take place. Fourth, communities or collectivities based on sustained identification or commitment give way to temporary associations based on shared actions; as a result social strategies based on the creation of sustained subjective investment or strong dispositions are largely outclassed by those more efficiently focused on producing specific effects at specific moments. Collectively, all four of these tendencies might be taken as a demonstrative of the moment when the dominant "logic" of culture begins to take on algorithmic or parametric forms.

In many ways, we have already encountered manifestations of several of these principles already in this text. Our discussion of changes in scientific epistemology and practice, in which the historical role of science as the privileged discourse of epistemology and representational validity largely takes a back seat to its pragmatic applications is certainly apposite. Similarly, our analysis of what we coded the technologic of contemporary capitalism, particularly in the ways that production and consumption have been connected in virtual feedback loops, would seem to fit fairly easily into this schema. We might additionally, however, detect parametric tendencies in a wide variety of additional domains.

Architecture

Indeed both "parametric" and "algorithmic" have become established terms in architecture, used to denote new design methods in which software programs and computing power are leveraged extensively to produce a kind of dynamic "program," as opposed to a static blueprint, for the construction of a physical structure. Most commonly, such programs work to maintain the integrity of a designed structure by automatically adjusting all elements of a design when another element is manually changed, but they are also used to generate multiple patterns and structural forms from defined criteria. If the central challenges of the practical nature of design—the literal production of a representation model for which a design will be based—used to be accuracy, fidelity, and iterability, algorithmic design methods call on architects to instead create ranges and a spectrum of acceptable effects (not only material "feasibility" but sizes, shapes, and accordances with other limitations of a design site or its environments). In this sense, such techniques instigate a radical shift in the representation economy of structural design, one in which, in the words of architectural historian Mario Carpo, architects are "leaving behind a universe of forms determined by exactly repeatable, visible imprints and moving toward a new visual environment dominated by exactly transmissable but invisible algorithms" (100). Much like advances in the ability to quantify spatial and visual units were key to the increased fidelity between the design and execution stages of an architectural structure, and increased capacities in calculating the dimensions of irregular structures and the "adequality" or interstitial space between unlike elements was key to the rise of the Baroque style, the contemporary use of parametrics might be taken as inaugurating a change of similar magnitude. Using computer-aided algorithmic techniques, it is possible not only to realize unusual visual or material forms in concrete structures, but also to set into motion the process of designing aleatory forms that are, sometimes quite literally, unknown or even inconceivable by the architect before being produced by a program.[8]

Here, again, it is worth noting how the introduction of automated techniques or the essential use of computing technologies into this field has had the *opposite* of the effects one might initially assume. In a strange sequence, the overtly mechanical or minimalist forms that we might intuitively associate with computer-aided architectural works is probably better represented by structures that predate them, the iconic buildings of modernist architecture and what cultural sociologist Mauro F. Guillén calls their "Taylorized beauty," a style invocative of both industrial environments and the scientific management of spatial efficiency. Algorithmic architecture, on the other hand, has most consistently been directed toward the realization of organic forms and ecological relationships, best represented not by clean geometries, but the "blobitecture" of creators like Greg Lynn, who uses a variety of computer-aided techniques to mimic what he calls an "anorganic vitalism" in both the design process and physical appearance of his works.

Still further, we might remark on the counterintuitive ways that the availability and ubiquity of massive computing power have also led to a scenario inverse to the supposed denigration or elimination of space and structure by networked communication technologies. Around the turn of the twentieth century what William J. Mitchell calls the "indispensable representational role" played by buildings seemed to be in the process of being gradually replaced by a variety of digital and virtual sites: company web presences for communication and e-commerce take over the representational role once played by iconic headquarters; elegantly organized databases take over from well-designed stacks in libraries; theatres give way to content collections of recordings and live-streamings (84). However, in the last decade or so, at least, we have seen a return of sorts to the production of iconic and monumental structures ostensibly made redundant by digital culture—even as many of them have been made possible by the same advances in computing power and availability. The difference, such as it is, might be that the symbolic value of such sites, as well as their role as indexing advances in technology and aesthetics, is based not on the industrial might or intrinsic artistic inventiveness of their creators and sponsors, but on the progression of computing power and algorithmic schemes for realizing them.[9]

(Mass) Media

The most visible impact of contemporary information technology and its attendant techniques on mass media, as a nonstop parade of "death of the newspaper" features have told us over the past several years, seems to have been one of displacement and fragmentation. If, as Jürgen Habermas among others has detailed, the progression of mass print media from the early 1800s onward has been one of consolidation—local and parochial print organs being absorbed or bought by national, and then international, firms—and a resulting heterogeneity of content and distancing from their core audience, movements from the late twentieth century onward might be seen as a total reversing of this trend.[10] We have, on the one hand, a disaggregation of the functions of centralized media like newspapers (whether in print or online) to a variety of new, typically localized or "local-centric" networked media—Craigslist takes over as the major avenue for classified ads, hyperlocal blogs take over local news coverage, and social media feeds become our new go-to resources for breaking news and trend-spotting. On the other hand, we have an integration of the formally "passive" reader or viewer into the production or content of media itself; the carefully curated and relatively rare "letter to the editor" gives way to the infinity of unfiltered comments on online news and blog features, and by-now quaint features like having audience members report the temperature in their backyard or submit on-air questions for television new anchors to field, is crowded out by the introduction of scrolling social media feeds and user-submitted video and audio of newsworthy events. Perhaps most notably, we have in many cases the total replacement of the professional journalist or commentator by "citizen journalists" or the crowdsourced coverage of events.

It is also worth remarking on how the parametric qualities of niche media have, perhaps paradoxically, been useful in maintaining particular equilibriums of opinion and disposition despite the ostensibly huge increase in perspectives and arguments of all strips in the mediasphere. If the primary negative effects of the centralization of media into large firms was that it tended to both restrict interaction between its producers and consumers, while also limiting the range of acceptable positions or opinions to those that were unlikely to offend a large audience—the qualities that, for good or ill, made media *mass* media—then it would seem that the vast proliferation of media sites and opportunities for interactivity would correct these problems, and thus perhaps even help inaugurate a new and more ameliorative modality of civil society or the public sphere. However, as you need go no further than the nearest comment thread on a controversial news story or partisan blog post to see, something like the opposite has been the case. Indeed, the unparalleled ability for audiences to search and access media created by individuals with which they already share political dispositions, as well as for aggregative technologies and media portals to "push" such media based on data collected about their users, has instead likely resulted in a decline in an individual's exposure to opposing viewpoints and of "rational-critical" communicative interaction between individuals with partisan divisions. Indeed, the realm of contemporary niche media looks much less like a global village or universal agora, and much more like an ever more intense balkanization of our political or ideological landscape.

Education

It might fair to say that education, at least university-level education of the past three to four decades, has been at the forefront of the integration of parametric techniques into praxis. Indeed, since the 1960s we have seen the introduction of a wide variety of techniques that orient pedagogical goals away from the teaching of discrete information to a variety of more generic capacities as well as the creation of a multitude of practical processes that attempt to microtarget and adapt to individual students learning strategies and deficits, and/or coordinate the educational experience around collaborative relationships between students. The emergence of process pedagogy in writing instruction, for instance, modeled a shift away from placing priority on a student's finished projects (the "product") and the evaluative and teaching method appropriate to that approach, replacing it with a focus on teaching discrete processes that form the smaller components of effective writing, and, subsequently, the evaluation of a student's work as a progression from the beginning to the end of the class (portfolio evaluation), and the integration of self-reflection on behalf of students as a method for making their writing processes apparent to both themselves and their instructors. To give just one more example, the increase in "problem-based learning," particularly in medical and engineering fields, served to redefine learning as a gaining of the flexible capacities to analyze and provide responses to loosely defined problems for which multiple solutions are likely effective. All of these processes are often performed in some way collaboratively between student groups, joining a variety of other practices—peer learning and peer evaluations—that have reconfigured teaching as the arrangement of relationships effective to learning rather than the explicitly hierarchical "transmission" of knowledge and skills from instructor to students.

To these examples we might note, yet again, the ways in which the actual introduction of information technology has had effects far from different from initial predictions. The big "innovation" for education promised by networked communication technologies and online environments was supposed to be, at least for teachers of an earlier vintage, the denigration of the importance of the school as a physical site and related enrichment and streamlining of existing capabilities for "distance education." The concern about these developments was that it would lead to a kind of formulaic approach to teaching, in many ways the very occlusion of the "best practices" referenced above that developed around student-centered "individualized" learning. Fast-forward to the present, and the fastest-growing area of technology in the classroom initiatives is the creation and use of "media rich" and "flexibly responsive" online environments—the variety of online portals through which students' performance on diagnostic exercises and other available data are used to custom design sequences of activities and exercises targeted at their particular strengths and weaknesses. If these systems sometimes still have the patina of what used to be called "skill and drill" approaches to teaching, they are also ones in which both sides of the equation—the measurement of a student's skills and their exercises they are asked to complete—are constantly being reconfigured in relation to one another and always around the individual student.

Finance

The dominance of financial speculation (investment driven not by analyses of existing underlying value but presumed future sales prices) with the increased "interconnectivity" of various stock markets and market sectors are often referenced as the defining factors of contemporary capitalism, with some intersection of the two often referenced as the key cause behind the recent global financial crisis. These two attributes themselves, however, are hardly new techniques in capital's history: indeed, one might say they are more or less what "made" capitalism into a recognizable economic system. Indeed, Giovanni Arrighi has detailed in his sprawling historical work *The Long Twentieth Century* how speculation's appearance at the center of investment strategies signals the start of something like a Nietzschean "eternal return" for state and then global capitalism since at least the fourteenth century, an indicator of systematic crisis that heralds the beginning of a new cycle of accumulation, expansion, and investment (214–238). Likewise, market interconnectivity, we might say, is a prerequisite for what any living person today would recognize as capitalism. We have records going back to at least the mid-1700s of the ways in which market failures and weaknesses lead to immediate changes in price and trading behavior in geographically separate trading spaces, and the increased connectivity of markets could be seen as one of the most reliable and consistent features of capitalism's history.[11]

Rather I take it that the truly novel characteristics of finance today, the changes that people are actually trying to refer to when "speculation" or "connectivity" are discussed in this fashion, are the increases in the magnitude and variety of market relationships and in the subjects of market speculation, as well as the ways these phenomena are themselves parametrically or algorithmically connected in contemporary financial exchanges. We can see this characteristic formally in the ways that modern trading is based not so much on calculations of underlying value or even presumed future value, but on the present and future state of a vast variety of associations between commodities, prices, and currencies—ones largely defined by the mechanisms of formal exchange itself. For one, there is a vast expansion of the kinds of things one can "speculate" on, including, in many ways, others' (or your own) speculations on the value of commodities; all investment trends toward being a virtual form of arbitrage, exchanges based on the relative values of different commodities and financial instruments. If stock exchange regulators of the late nineteenth century were disturbed by the idea that investors might be involved in speculation around agricultural products they do not actually plan to physically transfer, we can only imagine the their absolute bewilderment at the vast expansion of "financial products" in the late twentieth century to include percentages of the future profits of a city's parking meters or the bundling of mortgages into securities.[12]

As Daniel Buenza and David Stark write in their ethnographic study of an arbitrage firm, the modern day trading room "is an engine for generating equivalences": "traders locate value by making associations between particular properties or qualities of one security and those of other previously unrelated or tenuously related securities" (373, 376). As they go on to argue, arbitrage follows a logic and methodology far different from the "momentum trading" of the go-go 1990s, and in many ways the inverse of the corporate raiding strategies of the 1980s. While raiding strategies are based on carving up a company into assets and selling it piecemeal, arbitrageurs "carve up abstract qualities of a security":

> For example, they do not see Boeing Co. as a monolithic asset or property but as having several properties (traits, qualities) such as being a technology stock, an aviation stock, a consumer-travel stock, an American stock, a stock that is included in a given index, and so on. Even more abstractionist, they attempt to isolate such qualities as the volatility of a security, or its liquidity, it convertability, its indexability, and so on. (376)

In other words, in the realm of arbitrage, it is not the ability to break down an asset into component parts that might somehow be more valuable than their sum, but the ability to identify and manipulate a wide variety of its connections and co-implications with other phenomena, the flexible management of associations whether than the efficient calculation of an underlying value.

Democracy

Or at least that aspect of democracy that is most immediate and "direct" to most of the world's population: voting and the vast apparatuses of campaigning and persuasive messaging surround them. As briefly touched on in the last chapter, there has been something of a merging or recursion of techniques surrounding political campaigning and more traditional or explicit forms of marketing around consumer products and services. If the disappointing contention that our exposure to the ideas and actions of politicians and candidates has looked more and more like entertainment media and advertising—the increased use of televisual and later media increasing politician's resemblance to actors on stage, and political ads taking on the kind of vulgar appeals once reserved for selling used cars and high-end electronics—is by now a familiar one; perhaps what *is* new about this intersection how both increasingly traffic in data demography and niche-driven targeting of smaller groups with more specific similarities.

Although voters have long been targeted in line their presumed allegiance to particular "interest groups" (those invested in, for instance, income tax decreases, or against expanded international trade), in the last decade or so data-aggregating companies working for campaigns and political parties have combined computing power and the increasing availability of demographic and consumer information to provide their clients with a seemingly infinite number of cross-indexed voter niches. Such categories are not so much discovered as they are created by cross-referencing and compiling "psychographic" characteristics not necessarily associated with each other. One then develops ways to reach out to such groups to demonstrate how a candidate's (old or new) values mesh with theirs. As veteran republican campaigner Dan Schnur explains, although the general public might intuitively presume that campaign personnel develop a message and then test it to see how different segments of the electorate will react, the opposite is true of modern-day electioneering, which works by "identifying the various communities within the electorate and then shaping a message (or messages) that can reach each of these groups in the most beneficial manner" (360).

All of these phenomena, I take it, are manifestations of a parametric logic that has been spreading across contemporary culture over the last few decades. Perhaps what is most striking about them, as alluded to above, is the ways in which they seem to have had quite different, even oppositional, effects than those most commonly associated with information technology and the statistical-analytical and data-mining techniques associated with computing. If our default imagination of both—particularly as they began to become ubiquitous aspects of social life—has trended toward fears of their excluding, homogenizing, or depersonalizing tendencies, recent history has largely proven the opposite to be the case. Insofar as such techniques and technologies form a crucial part of contemporary political economy and cultural life, they have done so by maximizing the opportunities for inclusion, heterogeneity, and the "personalization" of everything from shopping recommendations, to medical treatment, to political messaging. Far from reducing everything through some inhuman logic or reduction or standardization, we might say in fact that contemporary technics has been very much premised on what Nietzsche called those "human, all too human" capacities toward "the overturning of habitual evaluations and valued habits," the constant transformation and creation of new conceptual categories and associations that occupies a wide range of economic, aesthetic, and otherwise social life today (*Human* 5).

All of this, of course, still leaves us with a series of pressing questions, many of which might be taken as amplified versions of the same monitory concerns that drove Heidegger's late-career writings on technics and the dissolution of metaphysics into cybernetics. Even if the spirit of technology today is very much the inverse of our usual "mechanistic" or reductionist understandings of the overcoding of the natural by the artificial, does that not suggest a level of appropriation within technical systems far more extreme than even Heidegger feared, one in which the "nature of nature" and of organic life is itself subsumed and simulated by media and technology? Is the tendency of capitalism and social power to integrate and draw value from a much broader range of identities and behaviors merely a symptom of its total saturation of cultural realms, this logic of inclusion much harder to disrupt than earlier forms of power that functioned via the exclusion or isolation of the "abnormal" and abject? Does this latter point, in particular, testify to the obsolescence of what we came to call "theory," the "intellectual" component of progressive or resistant social movements? In short, it is tempting to conclude that the eclipse of representation "itself" as a dominant mode of cultural production and contemporary social life is, if anything, only a more pointed or painful symptom of Heidegger's "darkening of the world"—and one even less available to be illuminated by traditional modes of critique, or by comparisons with an earlier era were experience and individuality could be seen as more immediate or "authentic" (*Introduction* 47).

There is much to support such a conclusion, particularly as long as we continue to presume that our best methods for understanding and intervening within social power comes by positioning ourselves as "oppositional" to its functions or "above" or outside its structures of reason or representation. However, allow me to give the hermeneutic circle one more spin here and suggest a different conclusion: that the convergence of technics and media today, the gradual transition of dominant logics of culture from the form of *logos* and metaphysics to structures and arrangements more traditionally restricted to realms of *techne* and rhetoric, make such distinctions beside the point. In other words, if the previously separate logics of the organic and artificial, the authentic and the ideological, of domination and of resistance, have reached a kind of null parity in the saturation of culture by cybernetics or parametrics, then we will have to do some very hard thinking about what possibilities exist within this new realm—how it might be oriented to produce different effects, and how we might fashion effective rhetorical, political, and ethical strategies in response to them. In large part, the rest of this book is dedicated to precisely these questions—diagramming the ways in which our understanding and practice of persuasion and motivation, power, and ethics could and should be rethought in response this new mode of culture. Right now though, and by way of concluding this chapter, I want to answer what I will argue might actually be the easiest of the questions listed above—the status and potential of "theory" today.

THEORIES OF TECHNIQUES, TECHNIQUES OF THEORY

To reprise one of the situations with which this chapter began, recall that much of the recent participants in the "death of theory" discourse of the past decade or so read theory's presumed demise as portent of, or synec-doche for, a greater shift in contemporary social power. What we might call the *theory of* "the death of theory" is that its failure to be a vital force in influencing contemporary culture is the canary in the coal mine that signals more far-reaching and dangerous cultural trends, namely, the mor-bidity of "critique" as a viable political tool and the co-option of theory's traditional methods of complication, contextualization, and skepticism by precisely the kind of retrogressive social actors that used be the subject of such analyses.

Thus we might say that the questioning of the validity of critical the-ory is not an attempt on retroactively indict the operation as it has been practiced across history of intellectual and academic thought, but rather to acknowledge its present collusion with, or dissolution within, the broader context of social power, its having become of a piece with the forces of domination or subjection that it was intended to combat. Such a conten-tion, however, might itself prompt a certain limited rereading of the recent history of critical theory or of its necessary or constitutive qualities. We might, as Hardt and Negri suggest, take the popularity of postmodern the-ories of difference inside and beyond the domain of critical theory as merely a "symptom of passage" into the times in which flexibility in identity and behavior becomes a crucial vector of value production under international capitalism. Or we might alternately simply presume that such contempo-rary conditions deprive critical theory of what Wendy Brown (channeling Nietzsche) refers to as its constitutive quality of being "untimely"—as a hesitating or oppositional response to current conditions and crises, one that seeks to produce "a rupture of temporal continuity, which is at the same time a rupture in the political imaginary, a rupture in a collective self-understanding dependent on the continuity of certain practices" ("Untime-liness" 7). In the final analysis, though, whether we read it diagnostically as a symptom of passage or as just a sign of the times, the conclusion seems to be the same: critical theory has lost its power as its methods have become at home within the cultural and political systems that are supposed to be its "object" of critique.

However, and perhaps in homage to the traditional conception of critical theory as a force that "slows down" lockstep conceptual linkages or even inverts (in the style of Marx and Adorno) conventional wisdom, I will sug-gest that this is too hasty of a conclusion, and that, in fact, the opposite may be the case. We might consider this along two related points.

First, it seems rather odd, at least to me, that the absorption of the tra-ditional tools of critical theory into social power "itself" would be read as the sign of its *failure*. Indeed, it would seem that, instead, the integration of critical-theoretical touchstones—the identification of potential biases in logic or representation, the challenging of binary or reductionist cat-egories, and, perhaps most importantly, the emphasis on the importance of culture in shaping human reason, disposition, or self-identification—is a ringing endorsement of critical theory's *success*. More specifically, if one of the original notions of critical theory was its thematizing of capitalism and institutional governance as prizing effectivity and efficiency over everything else, then it would seem axiomatic that the appropriation of critical theory into these domains is an undeniable testimony to the "power" of critical theory.

Perhaps it is easier to see things this way from the perspective of the appropriator rather than the appropriated. Consider, for instance, a recent interview with Andrew Breitbart, the publisher and multimedia mogul who—from his pioneering of social and "user-driven" media for conservative causes, to his role in fomenting the so-called Tea Party—has been involved with virtually every version of the "theft" of critical-theoretical practices by conservatives. Speaking with the *New Yorker*, Brietbart goes into great detail about the initial confusion he experienced when exposed to critical theory in his American Studies courses as a Tulane undergrad—"What the fuck are these people talking about? I don't understand what this deconstructive semiotic bullshit is. Who the fuck is Michel Foucault?"—before he realized the "real" lesson of critical theory (Mead). Arguing for a fairly straight line running from the emigration of Frankfurt School intellectuals to the U.S. in the 1930s to the election of the "radical" Barack Obama to the U.S. presidency over a half-century later, Breitbart emphasizes the revelatory effects of understanding the central subtext of critical theory—that culture can shape politics and opinion—as well as of the by then noticeable legacy of critical theory as a pedagogical or pragmatic undertaking, the circulation of *theories about culture* into culture, education, and the arts. For Breitbart, the American left has historically been more capable at using this strategy to their advantage, putting American conservatives at an extreme disadvantage:

> The left is smart enough to understand that the way to change a political system is through its cultural systems. So you look at the conservative movement—working the levers of power, creating think tanks, and trying to get people elected in different places—while the left is taking over Hollywood, the music industry, the churches. They did it through academia; they did it with K–12.

Whether or not one wants to read Breitbart's comments as part of a larger narrative in populist conservatism that seeks to flip the script of liberal critiques of power by positioning "liberal elites" as the true taste-makers and ideological programmers of present-day society, I take it the effect is largely the same: whether driven by the "actual" success of such strategies in reshaping political economy, or in the actual performance of its effectiveness in being appropriated by right-wing pundits such as Brietbart, the strategies of critical theory are far from moribund ones in contemporary culture.

Additionally, and to move on to our second point, I take it that if rehearsing the reading of a historical ontology by reference to changes in techniques rather than reason, representation, or technology as abstracted properties has anything to teach us, it is this: that (critical) theory has always been present or effectively operated as a technique, as a persuasive force for changing individual's dispositions and conceptual frameworks, rather than a negative "exposure" or demystification of reason or representation or the revelation of the "reality" hidden behind social conditioning. Indeed, even a cursory rereading of the history of critical theory along the lines of what it *does* rather than what it *claims* would reveal that this has always been what we might call the "function" of critical theory.[13] To somehow disparage the enterprise because it seems to have been successful in this function is to stubbornly cling to the Heideggerian separation of thought and technics, the insistence that even attempting to measure critical thought's value in practical terms is paradoxical, that that the idea that one might evaluate it "according to the standards that one would otherwise employ to judge the utility of bicycles" is a blasphemy against its very nature.

Finally, then, we might be able to give a more affirmative reading to two of the memorable phrases about the value of critical theory, and the state of culture after cybernetics, with which we began this chapter. Foucault's contention that theory can work to "show people that they are freer than the feel" might be more accurate than ever, as long as we alter the first part to read that theory's aim is to *convince* people that they are freer than the really are, not to somehow "free them" through demystifying the ideological frameworks or "social constructions" that are somehow limiting their behavior, but to actively persuade individuals to practice or perform actions and effects that they may have believed to have been inaccessible. And in this sense, then, Heidegger may indeed be right that today's age is marked by a "will to will," but we would also have to reevaluate or reinterpret that force as a very important one, indeed. To convince someone to will something, to persuade individuals that they can have an intense investment or disposition toward particular goals or results is no small feat. And in both of these endeavors, critical theory has proven to be a quite effective technique.

3 Rhetoric in the Age of Intelligent Machines

Burke on Affect and Persuasion after Cybernetics

> Indeed, always beneath the dance of words there will be the dance of bodies, the mimetic symbol-system that all these animals will come close to having in common, though their sedentary ways of living will cause them to forget it, like persons still quite young, come in time to forget the language of their childhood, the language most profoundly persuasive of all. But talk of the dance, and its body-language, brings us to exactly the next step in our unfolding.
>
> —Kenneth Burke, *The Rhetoric of Religion* (288)

THREE FOLDS

Around 2003 a participant identified as "Kenneth Burke" began making the rounds in chatrooms and other online environments, announcing itself with a morbid invocation of one of Burke's central concepts: "Hello I am Kenneth Burke; even though I have been dead for a few years, I would like to enter this conversation." This eerie salutation from the digital beyond is not an instance of online haunting, but the "Kenneth Burke chatbot"—a program with enough language-recognition coding and stored Burke quotations to converse in a combination of case-specific response and Burkeisms.

Although Burkebot's salutation lends a certain literality (however sinister) to the "unending conversation"—Burke's canonical metaphor for the continuity of human discourse—Burke's digital reincarnation as symbol-using code-script simultaneously undermines one of his more protracted interests: a career-long aversion to the becoming-machine of humans (and vice versa) that he saw occurring in registers both literal (e.g., the increasingly uncreative and mechanistic nature of human labor) and theoretical (through the emerging control sciences of cybernetics and its interdisciplinary inquiry into "control and communication in the animal and machine").[1] Burke's struggle against mechanistic representations of human bodies and human cognitive capacities begins in his earliest works and becomes increasingly proleptic throughout his career in contention with material advances in technology and speculative advances in cybernetic theorizing.

Human exceptionality in Burke's thinking continually narrowed in the age of intelligent machines, requiring defenses that shift from questions of epistemology to vectors of response:

> As regards our basic Dramatistic distinction, "Things move, persons act," the person who designs a computing device would be *acting*, whereas the device would be but going through whatever sheer *motions* its design makes possible. These motions could also be so utilized as to function like a voice in a dialogue. For instance, when you weigh something, it is as though you asked the scales, "How much does this weigh?" and they "answered," though they would have given the same "answer" if something of the same weight had happened to fall upon the scales, and no one happened to be "asking" any question at all. The fact that a machine can be made to function *like* a participant in a human dialogue does not require us to treat the two kinds of behavior as identical. ("Mind" 64)

Burkebot's mastery of symbolicity mocks this initial distinction (and through Burke's own words, no less). Unlike Burke's hypothetical scale, it differentiates between "intentional" and "accidental" stimuli and the question of whether we should "treat" its dialogue like that of a human participant becomes a non sequitur insofar as this difference reaches high levels of indiscernability in an electronic format. Similarly, the relative autonomy of Burkebot troubles the clean distinction of the agency of the operator versus the machine. Programmer, program, and source merge as Burke becomes sampled and Burkebot attains the agency of the DJ—sampling Burke into its emissions drawn from and reinserted inside symbolic environments.

Burke made allowances for such advances, admitting that cybernetic innovations might eventually require a complete rethinking of Dramatism; although, in opposition to both Darwin and cybernetics, Burke will argue that the difference between humans, animals, and machines is one of kind and not degree, the biological and the mechanical are configured by Burke into a spectrum of qualitative difference: "man differs qualitatively from other animals since they are too poor in symbolicity, just as man differs qualitatively from his machines, since these man-made caricatures of man are too poor in animality" ("Mind" 64).[2] Burke's (and cybernetics') three subjects become less static beings than the loci of capacities differing in their potentiality, vectors, and application. The implications of this position—the qualitative location of the human between and betwixt the animal and the machine—are foregrounded in an appendix to Burke's "Definition of Man":

> The idealizing of man as a species of machine has again gained considerable popularity, owing to the great advances in automation and 'sophisticated' computers. But such things are obviously inadequate as models, since, not being biological organisms, machines lack the

capacity for pleasure and pain (to say nothing of such subtler affective states as malice, envy, amusement, condescension, friendliness, sentimentality, embarrassment, etc., *ad nauseum*). (23)

Here Burke discovers human difference beyond the realm of symbolicity and signification; the affective states and capacities he introduces simultaneously hail capacities not only distinct from machinic intelligence but also outside of (or at least prior to) human rationality, referentiality, and agency. The other half of this cycle closes in "Mind, Body, and the Unconscious," where Burke writes that "a *conditioned animal*" provides a better model than the computer for "reductive interpretations" of the human, but that this animalistic parallel fails as well due to weaknesses in "the ways of smiling and laughing" (64). As per the epigraph above, in his work Burke continually moves beneath the dance of words to the dance of bodies; human difference emerges in organic ruptures such as laughter, which Bergson reminds us is not only a pivotal capacity separating humans and machines but the appropriate human response to the denigration of this separation— "*Something mechanical encrusted on the living*" (*Laughter* 39). Beyond the machine, laughter simultaneously trumps the reduction of humans animalistically—animals never get the joke—and takes us beyond human cognition or epistemology ("what's so funny?") to affective response.

This attention to aspects of human embodiment and their affective and asignifying corollaries run consistently through Burke's corpus, playing a central role from his earliest writings to later essays written in the 1960s through 1980s. My wager in this chapter is that assaying these movements in Burke's thought, engagements that flow between the seemingly oppositional ecologies of rhetoric and physiology and consistently put Burke in contact with the coterminous development of first-wave cybernetics, has much to add this book's attempts to thematize persuasion "after cybernetics"—the ways in which computing technologies and new media have changed the bases of public persuasion.

Indeed, Burke is a particularly salutary figure for the productive encounter of rhetoric and cybernetics for reasons in addition to the parallel development of his theorizing and that of first-wave cybernetics; beyond the disciplines' mutual interest in technologies (both mechanical and discursive) of "control" and "communication," Burke's early work on rhetoric and aesthetic theory foregrounds the more sophisticated intersections I will map in the remainder of this chapter, engagements perhaps best characterized by instances of another recurring trope in Burke's works—foldings and unfoldings: (1) a folding of modernist and postmodernist thought developing in the wake of informatic models of human consciousness that Eve Kosofsky Sedgwick and Adam Frank have termed "the cybernetic fold"; (2) a folding of "form" or structure that Burke locates in not only aesthetic and rhetorical tropes but also the physiological responses and rhythms of the human body; (3) and a folding of thought and agency that gets off of the ground in the writings of Nietzsche and Bergson and is updated by Burke to account for how interactions between our affective dispositions and social forces shape contemporary persuasion. Respectively, these three folds can be read as *structural* (delimiting both the general scope of rhetoric and the individual boundaries of the many concepts and processes Burke will invent), *transvaluative* (redesigning the relative influence and connections between forces and agencies at play in these processes), and *strategic* (discrete tactics for producing certain effects within particular environmental conditions). All three operate by a logic of "folding"—the limited manipulation of a particular field, one that maintains a given totality while creatively altering flexible aspects of it, such as an origami master's limited manipulation of sheets of paper.

We might say that the first of these folds begins after behaviorism and somewhere between dialectic and infinity. Though Burke was consistent and incessant about his disaffection for the emerging field of cybernetics, he shared many of the same concerns and points of departure with this interdisciplinary inquiry: an attempt to move beyond the stimulus/response pairing of behaviorism without recourse to depth psychology; a resultant concentration on internal rather than external mechanisms of persuasion and control; and an intense focus on affinities and divergences between human and machinic cognition and symbolicity.[3] Writing in reference to American cyberneticist Sylvan S. Tomkins, Sedgwick and Frank detail the results of such a "cybernetic fold": modes of theorizing inspired by "the moment when scientists' understanding of the brain and other life processes is marked by the concept, the possibility, the *imminence*, of powerful computers" and based on a model constituted by components greater than two but less than infinity (*infinity>n>2*) (508). This model, today most commonly associated with structuralism, is fundamentally pragmatic in its aim, guaranteeing that any theoretical taxonomy cannot be reduced to a vulgar dialectic (an "either/or"), but is also delimited into finite and identifiable categories. Although the production of further categories is possible, the *infinity>n>2* calculus demands that a theorist work immanently within an established system (though aware of its limitations) toward a given aim. For Sedgwick and Frank, this division of totality into finite components, a hallmark of structuralist thought, disappears in critical theory's "sleek trajectory into poststructuralism" through the latter's emphasis on social and discursive construction, denigration of the biological, and reliance on irreducible multiplicities.

One of Tomkins' most famous applications of the *infinity>n>2* formula is helpful here in foregrounding both Burke's engagement with this mode of cybernetics theory and its implications for his revaluation of rhetoric:

> We have assumed that the major motives consist of eight primary affects: interest, enjoyment, surprise, distress, fear, shame, contempt, and anger . . . These facial affective responses we assume are

controlled by innate affect programs which are inherited as a sub-cortical structure . . . These innate responses are later transformed in various ways through learning, but there is always a continuing openness to activation of the innate pattern of response. (Thomkins and MCarter 261, 219)

Burke referred to the first assumption here (the parsing of motive into eight primary affects) as an act of "scope and reduction," a strategy he introduces in direct opposition to behaviorism (*Grammar* 59). Burke's work is replete with such constructions; the logic of scope and reduction undergirds not only the five (or six) components of Dramatism, but also Burke's four "Master Tropes," eight varieties of the unconscious, five levels of linguistic significance, four primary mechanisms of historical change, and so on. The key upshot of such a formation is not its epistemic value (as Burke consistently reminds us), but its pragmatic potential. Structurally, it facilitates not so much epistemological resolution ("so that's what it is!") and the end of investigation, but a dynamic potentiality for functional application ("so that's what it does . . .") and the beginning of practice. In other words, the point of such an operation is not the exposure of a certain mechanism, but the potential that this mechanism might be manipulated through practice.[4]

Tomkins' statement is additionally helpful in articulating two folds also implicated in both cybernetics and the engagement Burke stages between rhetoric and human bodily dispositions, but outside of Sedgwick and Frank's formulation. Tomkins' second assumption approaches structure on another level—sub-cortical structure—but to this we might add many other biological systems and networks implicated in a similar way by both cybernetic research and Burke's interaction with this emerging field: neural nets, protoplasm, psychogenic and somatic response, and so on. These phenomena not only maintain the totality being divided through cybernetic folding—the "permanence" of Burke's *Permanence and Change*—but also locate originary (though, as we shall see, highly differentiable) capacities for response. These capacities, having developed ontogenetically and phylogenetically, run by the logic of Tomkins' embodied "affect programs" rather than strict rationality. In *Counter-Statement*, Burke finds these capacities rhetorically through the persistence and robustness of symbolic tropes and arrangements, and aesthetically through the artist's manipulation of "blood, brains, heart, and bowels" by inducing affective responses in an audience (36).

Tomkins' third and final assumption—the potential of *transforming* these affective responses—both reaffirms the permanence of these programs and introduces the possibility of their alteration. Here "program" moves from noun to verb, as the possibility for difference that always haunts repetition is mobilized and the body becomes not so much an inscribed site of actualization but a location for experimentation and transformation. Though, as mentioned earlier, contemporary readings of cybernetics often focus exclusively on the movement's impulse toward disembodiment and the creation of artificial intelligence, the movement was additionally (if not equally) committed to how cybernetic research might aid in transforming human perception and response. Not surprisingly—as a concomitant to the earlier folds mentioned above and as a general production of the impulse to diagnose the affinities and divergences between the human, the animal, and the machine—this impulse also traverses Burke's project in revaluating rhetoric, threading throughout his corpus but finding its most explicit formulation in his "Metabiological" concept of human communication and "Perspective by Incongruity," Burke's Nietzschean/Bergsonian method of breaking affective habitation.

The following takes up this tripartite (*infinity>n>2*) logic of structure (form), affect, and transformation in reference to Burke's early works *Counter-Statement* and *Permanence and Change*. This itinerary will ultimately fold rhetoric into the biological and mechanical (and vice versa) through these three smaller folds intersecting cybernetics and Burke's early writings; though more immediately recognizable to the former than the latter, folding becomes the practice of both the origami master and the rhetorician.

ONCE MORE WITH FEELING: REPETITION, AFFECT, AND PSYCHOLOGY OF FORM

Speaking nearly a half-century after the book's publication, Burke emphasizes the relative originality of his project in *Counter-Statement*: "I started from poetry and drama whereas most of such speculation starts from questions of truth and falsity, problems of knowledge. I started out with other words for beauty" ("Counter-Gridlock" 374). In *other words*, Burke responds to a world gone suddenly informatic—networked and linked by mass communication and rapidly proliferating media technologies—by dislocating knowledge; rather than entering epistemological territory, Burke shifts focus to aesthetics, rhetoric and their corollaries of structure, form, and feeling. As Burke explains in his response to Granville Hicks' critique of *Counter-Statement*, the imbrication of rhetoric and aesthetics offered in the essays "Psychology and Form" and "The Poetic Process" is focused on the effects of language and art and the responses they elicit rather than on judgment of these effects or prescriptions for the "proper" uses of art.[5] Burke is initially concerned not with

> *what effects should be produced*, but *how effects are produced*. In discussing the processes of walking, one must avoid any judgment as to whether a man should walk north or south. A moral imperative is not proper to a rhetoric, any more than the study of the mechanics of a motor equips us to decide whether motors should be used for warfare or trade.

Pushing beyond the normative models of both Marxism and Psychoanalysis Burke gets thoroughly physical. Art becomes "a coercive force in itself," unsubsumable to both superstructure and sublimation. As such the crucial site of persuasion emerges neither as the unconscious or false consciousness, but the corporeal.

As a unit, *Counter-Statement* is at the same time both a description and performance of the logic of form Burke is surveying. Burke's argument proceeds through a series of progressive divisions between the types of effects produced by aesthetic and persuasive discourses centered not so much on discrete works or artists as on the tropic forms they share both with each other and the processes of human physiology. By the conclusion of *Counter-Statement*, Burke's concepts are themselves divided and configured as tropes, recipes for the creation of particular effects in an audience (in the essays "Lexicon Rhetoricae" and "Applications of Terminology"). One of the first of these divisions emerges in Burke's distinguishing between two forces operating in any given communicative representation: a "psychology of information" manifested in the content of communication and focused on the transmission of signification from its producer, and a "psychology of form" contained in the expressive structure of communication and actualized through the effect it produces on the receiver. Ecologically, the two psychologies are engaged in a zero-sum game: "The hypertrophy of the psychology of information is accompanied by the corresponding atrophy of the psychology of form" (33). Discretely, however, the two exist differentially in any given representation. Referencing a satirical conflation of Cézanne's painting of trees with a forestry bulletin, Burke asks, "Yet are not Cézanne's landscapes themselves tainted with the psychology of information? Has he not, by perception, pointed out how one object lies against another, indicated what takes place between two colors?" The viewer cannot help but receive content from the painting; its image automatically hails spatial and temporal resolution, the reception of some cognitive signification even if it is only the knowledge of "how one object lies against one another." And yet the real force of the image resides not just automatically but autonomically in the affective, embodied response it provokes separate from its content or ostensible subject, "what goes on in the eye rather than on the tree" (32).

Such responses provide Burke a crucial assay through which to foreground the immediacy of aesthetic production, its seductive power to affect an audience or appeal through a process neither explicitly cognitive nor conscious. As it did for Nietzsche, the form of aesthetic production becomes for Burke a privileged site for troubling distinctions between solitary subjectivities, as, on the one hand, aesthetic forces are defined by their power to dislocate or distract our sense of self and familiar relations to our environments. Instead, they actualize capacities for response shared by a human multiplicity, ones that exert their own agencies often regardless of human decision:

A capacity to function in a certain way . . . is not merely something which lies on a shelf until used—a capacity to function in a certain way is an obligation so to function. Even those "universal experiences" which the reader's particular patterns of experience happen to slight are in a sense "candidates"—they await with some aggression their chance of being brought into play. (155)

On the other hand, Burke, arguing against prevailing modes of criticism that focus on the consistency of an author's corpus or the location of its work within a particular social or political program, continually under-scores the primary logic of aesthetic production as being located in the desires of the audience and their ability to appeal to affective capacities for response, forces he consistently codes as rhetorical.[6] For these reasons, Burke is consistently drawn to artists such as Flaubert, who, as shown in two letters excerpted by Burke in *Counter-Statement*, explicitly attended to the importance of form over content:

What seems beautiful to me, what I should most like to do, would be a book about nothing, a book without any exterior tie, but sustained by the internal force of style . . . a book which would have almost no sub-ject, or at least in which the subject would be almost invisible, if that is possible . . . I remember what poundings of the heart I experienced, what keen pleasure I felt on beholding a bare wall of the Acropolis . . . *Eh bien*! I wonder if a book, independently of what it says, might produce the same effect? In the precision of its groupings, the rarity of its materials, the unction, the general harmony—is there not some intrinsic value here . . . ? (6–7)[7]

The power of such a heterodox desire relies not so much in the autonomy of the work of art, a vector that might make it more easily subsumable to modernist form*alism*, but its attempt to build a work around the production of certain effects as embodied in a form rather than its discursive content or commentary. Such a form, Burke writes, is what constitutes the affective appeal of art: "One can repeat with pleasure a jingle from Mother Goose, where the formality is obvious, yet one may have no interest whatsoever in memorizing a psychological analysis in a fiction" (145). A nonsense rhyme, a bare wall, a book about nothing: all evacuate the signification or informa-tion of content to foreground the surfaces and structures of form. As they provoke our autonomic and automatic responses, they trace our relations to animals and machines based not on a kinship of identity, but an intersec-tion of shared capacities of response.

Attending to such forces was central to the polemical import of Burke's early work; the essays that compose *Counter-Statement* are meant to be first and foremost *counter*, contra to prevailing modes of interpretation of the time that focused on the facticity or meaning of expressive content and hailing an objective or disinterested perspective on aesthetic phenomenon. Here we might see Burke as anticipating in many ways more recent trends in cultural theory in the humanities and social sciences through which affect has emerged as an important counterpoint to conceptions of human experience as fundamentally shaped by modes of representation. As Massumi suggests, mechanisms of affect are neglected by what he refers to as "dominant" strains of social constructivism that argue "everything, including nature, is constructed in discourse" (*Parables* 38). Such a consensus has a serious impact on not only cultural criticism, but also critical conceptions of human bodies and human subjects: "The classical definition of the human as the rational animal returns in new permutation: the human as the chattering animal. Only the animal is bracketed: the human as the chattering of culture."[8] Recognizing affective forces and relationships then becomes vital to a necessary and "serious reworking" of the concepts of nature and culture, one that might express "the irreducible *alterity* of the nonhuman in and through its active *connection* to the human and vice-versa" (39).

Burke engages this latter objective in *Permanence and Change* (a point I will return to below), but as early as his writings in *Counter-Statement*, affective technologies were vital in articulating a new theorization of the relationship between rhetoric and the body. Burke locates these affects in not only the visual but also the verbal, citing aphorism—which "satisfies without being functionally related to the context"—as an example amongst a multitude of rhetorical tropes recuperated or invented through his "Lexicon Rhetoricae" (34). The crucial element of Burke's distinction between information is the *affective mode* of communication that operates before the level of signification—that which is produced by an image or articulation but that remains fundamentally unrepresented and unarticulated. As Burke details, this intensity is written immediately *on* the body but cannot be written down immediately *by* the body; its operation is algorithmically complex insofar as it cannot be represented but only reperformed in its entirety: "To arouse the human potentiality to be aroused by the crescendo, I must produce some particular story embodying the crescendo . . . Here I have replaced the concept by a work of art illustrating it" (46). Affect remains distinct from both signification and irreducible to emotion—a latecomer that can only respond to, rather than mediate, affect: "the emotions cannot enjoy these forms . . . (naturally, since they are merely *the conditions of emotional response*) except in their concreteness, in the quasi-vitiating material incorporation, in their specification or individuation" (46–47).

I am interested here in Burke's early theorization of affect for two primary reasons. First, it occupies a conceptual plane negotiated contra to both informatic reductionism *and* the kind of overreaching social constructivism that Massumi so compellingly critiques. Burke's concentration on affect would (as noted above) continually reappear as a principle of differentiation between humans and "intelligent" machines. The disjunction of "information" and "form" foregrounds the importance of the latter in human physiology and human interaction over and against the former term, one popular in emerging cybernetic work of Burke's time. Take, for example, the following passage from Norbert Wiener's *The Human Use of Human Beings*:

> the many automata of the present age are coupled to the outside world both for the reception of impressions and for the performance of actions. They contain sense organs, effectors, and the equivalent of a nervous system to integrate the transfer of information for the one to the other. They lend themselves very well to description in physiological terms. It is scarcely a miracle that they can be subsumed under one theory with the mechanisms of physiology. (43)

Burke's distinction between "information" and "form" would recognize the pragmatic use of this conflation of the biological and mechanical, but foreground the physiological processes he will associate with affective relations as unique to human beings. However, Burke's concentration on human physiology in this process additionally attends to factors of human corporeality and their relation with nonhuman phenomena in a manner that resists recourse to social construction. Moreover, his treatment of affect emphasizes the mutual reductionism they perform in different registers: just as a vulgar social constructivism would reduce the nonhuman as a cultural or discursive construction of the human, a vulgar cybernetics would reduce the physiological operations of the human body to informatics; the key difference in these operations emerges in the former's impulse to the critical *deconstruction* of phenomena under review contra the latter's emphasis on the material *construction* of pragmatic technologies based on this principle of conflation.

Second, and in a similar vein, Burke's early treatment of affect is salutary for reconsidering contemporary critical work on affect and embodiment in both the discipline of rhetoric—where affect has both been hypostasized as a systematic structure of influence available to be decoded and resisted (thus occupying a role similar to that which "strong" conceptions of ideology used to play in critical theory)[9] and the asignifying and "non-rational" has been invoked generically to critique logocentric or consensus-based schools of rhetoric—as well as humanist approaches to cybernetics, where, as we say in Chapters 1 and 2, embodiment and affective processes have been pitted against popular conceptions of artificial intelligence or informatic subjectivities.[10]

Burke's relatively novel and nuanced engagement with affective phenomena arises from a close attention to the underlying structures and conditioning factors of these forces. Though he was following humanist approaches to cybernetic theories that uphold affective phenomena as processes that (currently) cannot be captured by high-technological simulations of human consciousness or intelligence (even as the resources we use to quantify these phenomena—PET scans, infra-red oculography, facial electromyography, etc.—are increasingly of this nature), Burke additionally attends to the fact that these forces contain a consistency that is oddly mechanical in nature. In this sense, the pragmatic potential for manipulating affective forces is foregrounded over its use as an "outside" to either traditional rhetoric or early cybernetic conceptions of informatics, and Burke appropriates key vectors from both disciplines in his endeavor. Principle to this conception is Burke's argument that affective forms develop and sustain through an imbrication of physiology and repetition, a sequence he will associate with the rhetorical technology of the trope: forms of language and expression unified by their structure rather than their content and that share structural affinities with biological and nonhuman processes. He writes once again in reference to the "crescendo" or "climax":

> there is also in the human brain the potentiality for reacting favorably to such a climactic arrangement. Over and over again in the history of art, different material has been arranged to embody the principle of the crescendo; and this must be so because we "think" in a crescendo, because it parallels certain psychic and physical structures which are at the roots of our experience. The accelerated motion of a falling body, the cycle of a storm, the procedure of the sexual act. (*Counter-Statement* 45)

Later, he argues for systolic and diastolic processes as the cause of predictable rhythm in aesthetic and rhetorical creations (140). However, repetition maintains the ubiquity of these forms, a point Burke makes in detailing the "contribution" of the psychology of form as distinguished from the psychology of information: "Truth in art is not the discovery of facts, not an addition to human knowledge in the scientific sense of the word. It is, rather, the exercise of human propriety, the formulation of symbols which rigidify our sense of poise and rhythm" (42). Music—compositions that maintain their integrity solely through the production of form and repetition—provide the ultimate example of the force working through form: "The reason music can stand repetition so much more than correspondingly good prose is the music, of all the arts, is by its nature least suited to the psychology of information, and has remained closer to the psychology of form" (34). Form cannot atrophy here, and the asignifying, repetitive quality of music entirely occludes intrusion by the psychology of information. Nevertheless, these forms themselves manifest a consistency, making them available for use and manipulation.

Through his first sustained encounter with rhetoric and the body in *Counter-Statement*, then, Burke's double gesture introduces the affective, asignifying components of rhetoric as well as the rhetoricized faculties of the body: the capacity to detect and respond to forms that emerge from the natural processes of the body and are maintained by repetition. And yet, repetition, by nature of its continual iteration, invites the possibility for difference. In fact, as Burke notes, "restatements with a difference" are not just a possible mutation of form, but constitute one of the most reliable and consistent existing forms. I take up the potential this formulation carries for transforming thought and perception later in this chapter, for now I want only to note that Burke's emphasis on the "conditioning" that instantiates affect and creates the structures of for leaves open the possibility for further "reconditioning."

A CLOSER LOOK: PERSPECTIVE BY INCONGRUITY

So what of this body then? In *Kenneth Burke and the Drama of Human Relations*, William Rueckert titles his chapter on *Counter-Statement* "Both/ And." For Rueckert, Burke's theory of form creates a "merge" between "essences" and "structures" best described by this term often associated with poststructuralist thinking, a relation later replaced by Burke with the superior formulation of dramatism. More recently, Bryan Crable—drawing on the same "progress narrative" of Burke's writings that informs Rueckert's analysis—conceives Burke's theorizing of embodiment in *Permanence and Change*'s Metabiology as a misstep to be corrected by an "either/or"— the dramatistic dialectic of the action/motion pairing. Such a polarity provides for Crable a tool for distinguishing how symbolic action, in the form of social construction, is routinely passed off as non-symbolic motion—the "natural" (134).

However, Burke's engagement with corporeality in *Permanence and Change* might pose questions more vital to contemporary rhetoric, ethics, and subjectivity than either dialectic or social construction. Metabiology auscultates a relation that is neither the "poststructuralist" "both/and" nor the dialectical "either/or." The cybernetic logic of $infinity>n>2$ moves from form to body (the only way form ever moves) in the interstice between *Counter-Statement* and *Permanence and Change*; the presumably strictly autopoetic structures of body and culture are radically linked in this movement, one we might term after Deleuze and Guattari a logic of "either, or, or . . ." (*Anti-Oedipus* 12). For Deleuze and Guattari, this formation hails not dialectic or endless convergence, but a dynamic set of linkages and disjunctions. Through Metabiology, this affiliation follows the logic of what scientists such as Eric D. Schnieder have argued is the organizing vector of both the "symbolic action" of organisms and the "nonsymbolic motion" or tornadoes and global weather patterns—the gradient. The relation between the body and the symbolic (or culture, or the socius, or rhetoric) emerges not as opposition or subsumption, but "a difference across a distance" (Margulis and Sagan 45).

If the folding of form that marks Burke's project in *Counter-Statement* was a reaction to an informatically networked world driven by a contemporary "cult of excessive technological 'progress,'" then Metabiology assays the possibility of a different, more organic, and ancient topological connectivity: "the entire attempt to distinguish between organism and environment is suspect" ("Biology" 17; *Permanence and Change* 232). Joining early cybernetic scholarship arguing for the necessity of viewing human actions and thought processes as networked within larger ecological systems, Burke's Metabiological perspective augments first the organic and then rhetorical ecologies deployed in this equation, linkages that were rapidly being crowded out by a concentration on the material and theoretical human/machine fusions of artificial intelligence and automata. Conceiving the world as a meshwork of bacteria and genes in addition to (and before) one of wires and circuits—an anticipation of what we would now call the Gaia hypothesis: earth as "cybernetic system"—Burke extrapolates two consequences from this conclusion (Lovelock 131). For one, the body becomes more ratio than closed system and divisions become the production of language and terminology rather than structure: "Is oxygen environmental or internal? Are the microscopic creatures in our blood stream separate from us or a part of us? They are members of a 'civic corporation' which we call the organism" (*Permanence and Change* 232). For Burke, the environment cannot be the "distinctly prior factor" in any analysis of human activity and behavior (233). Although we can discern "events manifesting sufficient individuality from our point of view to be classed as separate organisms," the "universal texture" of experience is composed of a dynamic interaction between the sentient, non-sentient, organic and inorganic components of the environment.

Second, Burke emphasizes the "*participant* aspect" of these actions and exchanges rather than their "*competitive* aspect" (266). The neo-Darwinian emphasis on agonistic becoming as the primal relation among and between species and their environments is replaced by an unfolding structure of cooperation, communication, and homeostasis: "Life, activity, cooperation, communication—they are identical" (236). Thus, "fitness" of species and ecological change emerge not from a capacity to endure or overtake but to productively *respond*. The key upshots of this formulation are its resistance to a rigid separation between participants in an ecology and its refusal to subsume the biological into a subject to be either cognitively interpreted or dismissed as a social construction. The biological and the affective emerge as phenomena to be interacted with, and the relation between the physical and the symbolic becomes much more complex than simple cognition.

Indeed, although Burke is willing to take this ecstatic physicality far enough to disperse the *cogito* by insisting we follow Bergson in thinking of the "mind-body" as a merge rather than separate entities, the (apparently) exo-biological communicative structures of symbolic culture prove a thornier issue (*Permanence and Change* 94; Crusius 97). As a microcosm of this process, Burke draws on the developing field of endocrinology and the example of psychogenic illness to diagnose the connection between enculturation and physiologic response that is at play, for instance, in the cocktail of dopamine and (sub-)socialization that gives meaning to a term such as "drug culture." Writing in the afterword to *Permanence and Change*, Burke retrospects on the Metabiological body–culture relationship undergirding such a process:

> The principle of individuation (which is grounded in the centrality of the nervous system) features a dualistic distinction between "primal, immediate" sensations that we experience in the realm of nonsymbolic motion and the vast "mediated" knowledge of "reality" acquired by the learning of symbol-systems (resources in the realm of symbolic action that give us access to wholly public modes of interpretation, orientation, and corresponding cultural relationships). (313)

In *Permanence and Change*, Burke has a multiplicity of names for such "modes of interpretation, orientation and corresponding cultural relationships" in both their positive and negative actualizations: trained incapacities, pieties, occupational psychoses, and so on— "complex interpretive networks" developed through habituation.

Cognitive neuroscientist Merlin Donald is helpful in unpacking the varying connections between the symbolic and physiological, the impulse behind Burke's project to determine human "symbolic behavior as grounded in biological conditions" (275). Donald describes the unique symbol-using capacity of humans as both a radical openness and dependence:

> The human brain is the only brain in the biosphere whose potential cannot be realized on its own. It needs to become part of a network before its design features can be expressed . . . The result is that we are plugged-in, as no other species before us. We depend heavily on culture for our development as conscious beings. And by exploiting this connection to the full, we have outdistanced our mammalian ancestors. (324)

But this capacity comes with a cost; the "hybrid brain" of humans makes the individual "distinct" but never "fully autonomous" from the symbolic structures of their culture (326). The fact that the "modes of interpretation" installed by the symbolic technologies of culture form through repetition and practice rather than conscious acceptance suggest that they similarly cannot be removed by a simple process of exposure or rejection. Rationally, there would seem to be no changing these structures.

Writing in the afterword to *Permanence and Change*, Burke approaches such an obstacle in foregrounding the primary component of Metabiology that he still finds appealing: "Above all I would cling to the notion that the concept of a biological 'method' suggests a useful way of avoiding the oversimplified reduction to a blunt choice between 'rational' and 'irrational'" (297). I take up Burke's refusal to reduce situations to this binary in the following, an impulse that finds its most productive manifestation in his Perspective by Incongruity—a singularizing technology for altering structures of interpretation by working by the same logic that creates and sustains these structures.

As indicated above, concentrating on Burke's treatment of the body and its relation to affect and the symbolic through Metabiology reveals a crucial period in his engagement with these topics, a moment when Burke paused from cataloguing the development and circulation of orientations and "spontaneously opted for the principle of *transformation*" (54). Early in *Permanence and Change*, Burke writes that "shifts in interpretation make for totally different pictures of reality, since they focus the attention upon different orders of relationship" (36). Although, as Burke documents relentlessly, we "learn to single out certain relationships in accordance with the particular linguistic texture in which we are born," there is still the potential to "manipulate this linguistic texture to formulate still other relationships." This possibility for manipulation is mobilized in Perspective by Incongruity, a concept Burke develops in reference to two theorists who also performed significant critiques of the division between rationality and irrationality and share his views concerning the differential relationship between affect and content: Bergson and Nietzsche.[11]

For Burke, it is Nietzsche, who "knew that probably every linkage was open to destruction by perspectives of a planned incongruity," that provides the most vital contribution to the process of "experimentally wrenching apart all those molecular combinations of adjective and noun, substantive and verb, which still remain with us" (91, 119). In her analysis of Burke's debt to Nietzsche, Debra Hawhee emphasizes how the latter's "perspectivalism" (and in its adaptation by Burke into Perspective by Incongruity) works by a logic of multiplication (the quantitative accumulation and application of different perspectives) and a concomitant critique of totalizing systems of knowledge. Although this is undoubtedly true—Nietzsche and Burke both state as much, and it fits with Burke's later comments in *Attitudes Towards History* on the effects of the device—I would like to foreground here the connections between Burke's concept and emerging cybernetic conceptions of subjectivity as well as a concomitant aspect of Perspective that is more closely tied to Burke's reading of Bergson rather than Nietzsche: the logic of reversal and inversion by which it operates.

As detailed above, although Burke dismissed the rigid equivalences between humans and machines articulated in popular debates over cybernetics, engaging this problematic was salutary for his theorization of human affect and agency and the impact of rhetorical technologies in producing and manipulating these structures—conceptions that troubled traditionally humanist conceptions of subjectivity and self-sovereignty even as it resisted reductively informatic depictions of the same. As such, the discipline was equally premised on a nuanced conception of the imbrication of the individual and her environment, one that drew on informatics and automata theory to detail the impact of ecology and interaction on human knowledge production and human perspective. Such a conception was at the forefront of often-neglected cybernetic work in the human sciences, articulated, for example, by anthropologist and cyberneticist Gregory Bateson in his essay "Cybernetics of 'Self,'" a consideration of the coincidence between "the theology of Alcoholics Anonymous" and the "epistemology of cybernetics" (309):

> Cybernetics . . . [recognizes] that the "self" as ordinarily understood is only a small part of a much larger trial-and-error system which does all the thinking, acting, and deciding. This system includes all the informational pathways which are relevant at any given moment to any given decision. The "self" is a false reification of an improperly delimited part of this much larger field of interlocking processes. Cybernetics also recognizes that two or more persons—any group of persons—may together form such a thinking-and-acting system. (331–332)

Burke's Perspective by Incongruity would draw vectors from the seemingly oppositional dynamic of self-direction and interdependence, conceptual tools circulating in early cybernetic theorizing and still very much at play in trailing edge artificial intelligence programs based on response mechanisms—such as Burkebot—by positioning would-be practitioners of Perspective as both subjects and objects in its process. In encouraging readers to "deliberately cultivate the use of contradictory concepts," Burke positions them as first self-directed decision-making subjects, and then responsive and responsible objects inside a greater affective and conceptual network. This planned encounter with the seemingly illogical would aim at disrupting established conceptual habituations and chronic modes of response by both drawing on and challenging the aspects of human knowledge production and cognition that form our deepest affinities with both non-human animal cognition and mechanical "intelligences." As such, it constitutes a rhetorically based practicum in encountering and cultivating alterity—training in how to conceive and think differently.

Bergson's writings on metaphor and perception would provide for Burke the core of the new "approach to reality" actualized in Perspective by Incongruity and the complex consideration of human agency undergirding it (95). For Burke, Bergson's application of metaphor provides a way of disjoining and conjoining substances by working on a level other than rationality: "Indeed, the metaphor always has about it precisely this revealing of hitherto unsuspected connectives . . . It appeals by exemplifying relationships between objects which our customarily rational vocabulary has ignored" (90). Bergson provides perhaps his most concise treatment of this process in *Laughter: An Essay on the Meaning of the Comic*, a work focusing on that rather irrational response that, as detailed at the beginning of this essay, Burke saw as constitutive of human difference and potentiality. For Bergson, the "incongruous" marks not solely a critique of rationality and a disruption of habituated linkages, but an ability of the human mind to be affected by a logic other than that of rationality:

> Such a proposition as the following: "My usual dress forms part of my body" is absurd in the eyes of reason. Yet imagination looks upon it as true. "A red nose is a painted nose," "A negro is a white man in disguise," are also absurd to the reason which rationalises; but they are gospel truths to pure imagination. So there is a logic of the imagination which is not the logic of reason, one which at times is even opposed to the latter,—with which, however, philosophy must reckon, not only in the study of the comic, but in every other investigation of the same kind. (42)

The human capacities to perceive and respond to the incongruous provided not only the structure of comedy, but also a vital port of entry into investigating how perception and thought in general are disciplined into a cycle of rationalization, critique, and judgment. Intervention into this process requires a "special effort" to forestall lockstep judgment by cultivating habits of the imagination rather than reason:

> In order to reconstruct this hidden logic, a special kind of effort is needed, by which the outer crust of carefully stratified judgments and firmly established ideas will be lifted, and we shall behold in the depths of our mind, like a sheet of subterranean water, the flow of an unbroken stream of images which pass from one into another. This interpenetration of images does not come about by chance. It obeys laws, or rather habits, which hold the same relation to imagination that logic does to thought. (42–43)

Burke's Perspective by Incongruity is essentially a technology for retraining human response drawing on the capacity to respond to (in)congruity Bergson elaborates above. It works by nature of both meanings of "trope": "trope" as designation for static structures of form and language counted on for producing a consistent response, and the "trope" of "troping"— an action that introduces juxtaposition into this first mechanism of conditioned response, thereby producing a differential relation between the original structure and its perception. As Nietzsche reminds us, all perspectives are only perspectives, and as Burke reminds us, all perspectives are in a sense already "perspectives by incongruity": metaphorical constructions linking divergent phenomena. However, Burke's methodology of planned incongruity works by the acceleration and deceleration of these already ubiquitous processes. On the one hand we have a *speeding up* of perspectivalism; Burke echoes Nietzsche in calling for the quantitative multiplication of perspectives and the concomitant devaluation of any one single "correct" perspective. On the other hand, we have a deceleration of the habituated reaction to perspectives: a replacement of the lockstep impulse to judgment with hesitation. Burke hails a qualitative (in addition to quantitative) operation on perspective; the complexity of incongruity creates a space for response rather than judgment or rationalization. In this sense, then, Burke gets us much closer to the forces shaping structures of thinking than any deliberation on the process could achieve. The realization that orientation and conditioning shapes perspective on a level beneath rationality—perhaps the most consistent topic in Burke's corpus—eventually led Burke away from the "debunking" of such structures to the pursuit of the most beneficial one. Among the various perspectives Burke cultivated in his writings, however, Metabiology emerges as singular in one crucial aspect: it marks a point in Burke's trajectory where his attempt to provide the most appropriate interpretive frame leads us instead to the creation of tools for cultivating singular and differentiating shifts in interpretation.

CODA: ON SEEING DIFFERENTLY

Above I've attempted to map out sites in Burke's early work where new ways of thinking about the human body and the correlations between the symbolic capacities of humans, animals, and machinery prompted reinvestigations of the connections between the body and rhetoric. In most cases this relation has foregrounded how corporeality affects rhetoric, but Burke describes in the afterword to *Attitudes Towards History* a very personal event during the construction of the Perspective by Incongruity section of *Permanence and Change* where rethinking rhetoric affected *his body*:

> And precisely then, at a time when I was focusing on the concept of "double vision" and as I began seeing the design of my whole project changing, the twist of vision became actual. On the road to go shopping, I saw two cars coming whereas I knew it was only one, looking double. I could see close up without the doubling, but the farther off things were, the wider apart the two images became. What was this? Cancer of the brain, perhaps . . . Yet so far, no diagnosis; nothing but plans for further and costlier examinations. (399)

True to his consistent emphasis on the effects of psychogenic illness, Burke performs a self-diagnosis linking his physical ailment to his mental state and prescribes a self-experiment in response:

> I have had several occasions to learn that, if we get involved enough in the using of words, the words in turn begin using us. "Inspiration" is an *honorific* word (thus dangerously deceptive) for a process

of self-hypnosis than can result from over-susceptibility to whatever terms one happens to be engrossed with. So I diagnosed the situation thus: When speculating on the resources of the term "double vision" at the same time that I was shifting my perspective on my own books on perspective, I began seeing double. So I worked tentatively on the assumption that I was subjecting myself to the magic of some obvious "let there be" equations. I clearly "solved" the dizzying formal problem thus brought to the fore when the Nietzschean theme of "transvaluation" in the middle section of *P&C* introduced what Trotskyites might call a variation on the theme of perpetual revolution. My recovery followed forthwith—and you can't imagine what a truly sybaritic delight it was, to look down the road and see just one car coming.

Burke's "double-vision" marks human susceptibility to persuasion and connection, a capacity his early works detail as operating through the body rather than through rational cognition (a process "doubly" emphasized by Burke's physical symptoms). Perspective by Incongruity emerges as the "perpetual revolution" against the negative effects of this condition, a sustained retraining of response that manipulates rather than blocks human capacities to be affected.

Burke's early writings—which I have tried to illustrate as dynamic attempts to fuse rhetoric and the corporeal and recognize the importance of affect and the arepresentational—occupy a tentative place in his canon. Rueckert speaks for many (though perhaps more explicitly than *any*) when referring to these works as a "stylistic and terminological underbrush," and "an irritation, a distraction, the rank growth of a fecund mind" (5). When these writings are not summarily dismissed, they typically (as for Rueckert) function largely as anticipation for Burke's later works, adolescent versions of his more domesticated concepts such as dramatism and more generically familiar methodologies such as dialectic. The recuperation of Burke as a proto-poststructuralist by many critics has also contributed to an emphasis on the later works and the (perhaps paradoxical) canonization of Burke as both our preeminent rhetorical humanist and most important postmodern rhetorician.

However, I would like to end here by suggesting that Burke's early writings might provide more powerful (and more timely) tools than either humanist rationality, dialectic, or social constructivism and that he concomitantly provides a new *telos* for contemporary rhetoric. Having essayed the power of form and habituated structure to shape perception through the rhetoric and aesthetics of *Counter-Statement* and the Metabiological frame of *Permanence and Change*, Burke invites us to get involved in these processes themselves rather then their critique. Working immanently *through* rather than *outside* of these productions, Burke prescribes the redirection rather than elimination of already existing (and unavoidable) structures of influence. Such a process instantiates not so much a resistance to the force behind orientation, but a productive modulation of these forces gained through an engagement with their effects rather than their meaning: rhetoric emerges not only as a technology for persuading others, but as a technology of the self used by rhetors to discipline and transform their own habits of response. Picture Burke looking down the highway and delighting in his perception of only one car: a movement from *multiplying* perspectives to *manipulating* perspective, from seeing doubly to seeing differently.

4 Any Number Can Play
Burroughs, Deleuze, and the Limits of Control

If the seventeenth and early eighteenth centuries are the age of clocks, and the later eighteenth and nineteenth centuries constitute the age of steam engines, the present time is the age of communication and control.

—Norbert Wiener, Cybernetics (38)

The old sovereign societies worked with simple machines, levers, pulleys, clocks; but recent disciplinary societies were equipped with thermodynamic machines presenting the passive danger of entropy and the active danger of sabotage; control societies function with a third generation of machines with information technology and computers, where passive danger is noise and the active, piracy and viral contamination.

—Gilles Deleuze, "Postscript on Control Societies" (180)

Enemy have [sic] two notable weaknesses:
1. No sense of humor. They simply don't get it.
2. They totally lack understanding of magic, and being totally oriented towards control, what they don't understand is a menace, to be destroyed by any means—consequently they tip their hand. They don't seem to care anymore—but famous last words: "We've got it made."

—William S. Burroughs, Last Words (15)

In the summer of 2002, Seattle news sources reported on an unusual business venture by two web entrepreneurs. From July 7 to 14, Scotty Weeks and Derrick Clark feigned homelessness in Victor Steinbrueck Park. Attired in "homeless clothes" (carefully selected from a thrift store) and bereft of all modern conveniences (save a digital camera and audio recorder), the duo slept on the streets, begged for spare change, and updated their website (homelessweek.com) daily from a nearby Internet café. Their objective, Clark explains, was to "learn a lesson and see the world with different eyes" (qtd. in Jenniges). Feigning homelessness is not an altogether new occupation—for instance, in 1987 television anchorwoman Pat Harper famously spent six days pretending to be a homeless on the streets of New York to produce the sweeps week news special "There but for the Grace of God," and diverse ministries have long offered this opportunity to their congregation in an effort to build sympathy for the homeless—but Weeks and Clark's venture is relatively unique for two reasons. One, there is an explicit element of self-promotion: although their press releases indicate a desire to gain "humility" and "empathy" through this experience, the clear emphasis on the pair's personalities and their other web-based ventures displayed on homelessweek.com severely undermines this professed altruism.[1] This point is reinforced by the second item that makes homelessweek.com an odd enterprise: the seemingly contradictory exploitation of the homeless on Weeks' personal website, where one can view "hidden camera" images of the homeless having sex and defecating accompanied with amusing captions and little mention or demonstration of "empathy" or "humility."[2]

Although the pair's interaction with the homeless may be rather unique due to the juxtaposition of their press releases' concern with humility and empathy and the explicit exploitation of the homeless through the hidden camera photos, their promotional ventures—selling the experience of "homelessness" and attempting to profit from media captures of "real," material homeless individuals—are not incomparable; both find affinities with other, more explicitly profit-driven undertakings. For instance, around the same time as the duo's experiment, a Dutch travel agency had been marketing the concept of "homelessness" through "homeless vacations" where work-weary Europeans can pay to beg for change on urban streets for a week. Also during the same time, the Las Vegas website Bumfights. com was enacting a more direct way of making a profit from actual homeless individuals by selling videotapes and Internet footage of homeless men and women fighting each other and engaging in dangerous stunts. Since it is not illegal to film brawls in Nevada, and the majority of their homeless "employees" gladly accept small amounts of food, clean clothes, or alcohol as payment, the producers of Bumfights.com had developed a low-risk and highly profitable business based on poverty. In response to those who critique the Bumfights enterprise, creator Ryan McPherson argued that their relationship with the homeless has been mutually beneficial and that the Bumfights producers are in fact some of the very few who paid attention to the homeless and tried to provide them an outlet: "We were the only ones there and we're the ones that are friends with a lot of these homeless people that we filmed" (qtd. in Kowsh).

I dwell briefly here on recent media and economic movements clustered around both material homeless individuals and the concept of homelessness itself because it seems to me a particularly exemplary site for analyzing the intersection of the parametric mode of contemporary culture that was detailed in Chapter 2 with what we have come to call, after Foucault, "social power" as the system of forces that in diffuse and often implicit ways structure contemporary political economy. More precisely, we might say that this intersection largely begins where Foucault's concept of disciplinary society leaves off and that the suturing of homelessness as both a spectacle and as "experiences" available for purchase is a particularly salient example of the ways in which contemporary media and technologies have shifted the locus of social power from the confined sites of identity formation and management that were key features of discipline, to those of larger media ecologies wherein human capacities of various types are shaped and circulated. While it might be said that the variety of arguments already made in this book have addressed significant aspects of such a transition, in this chapter I pursue a more specific thematization of the impact of cybernetic technologies and concepts in shaping social power. More specifically, I am interested in asking how we can or should define categories of "domination" and "resistance"—those two great buzzwords of cultural theory of the past few decades—in such a changed environment.

But first, and by way of a refresher, we might begin by considering a few ways in which the events with which we began this chapter present particularly poignant or material examples of two phenomena we have already had occasion to analyze in this text. First, these instances certainly provide yet a further demonstration of the increasing flexible basis of economic production and consumption marking the present. As Fredric Jameson suggests, such processes are currently marked not so much by the simple switch from the dominance of a product-based economy to a service-based one, but rather by the introduction of an entirely "new ontological and free-floating state" of economic production, one in which

> content, (to revert to Hegelian language) has definitively been suppressed in favour of the form, in which the inherent nature of the product becomes insignificant, a mere marketing pretext, while the goal of production no longer lies in any specific market, any specific sets of consumers or social and individual needs, but rather in its transformation into that element which by definition has no content or territory and indeed no use-value as such, namely money. ("Culture" 153)

In such a manifestation, economic production and profitability emerges not so much through the manufacture and marketing of particular products or even services in the traditional sense, but through a marketplace based on the creation and availability of new experiences and sensations— the thrills and risks of gambling and financial speculation, entry into the virtual adventures of video games and related media, the opportunity to "experience" homelessness.

Second, and with closer proximity to the subjects of our previous chapter, we might also suggest that such phenomena also emphasizes the ways in which affect or emotional empathy, as either an asocial or uncommodifiable space, or as a reliable source of collective empathy or sympathy as in what Honig calls "tragic" humanism, seems to have increasingly diminishing returns. Indeed, it would seem that, insofar as such affective ties were based at least in part in shared identities and relatively stable sites of collective experience, then the decline of disciplinary power implies a concomitant decline in affect as a strategy of resistance or of social amelioration. For instance, in assaying what she calls "sentimental politics," Lauren Berlant has powerfully underscored the ways that shared capacities for affective feeling, particularly those of pain and suffering, have traditionally been mobilized to foment social change:

> Sentimentality has long been the means by which the mass subaltern pain is advanced . . . to eradicate the pain those with power will do whatever is necessary to return the nation once more to its legitimately utopian odor. Identification with pain, a universal true feeling, then leads to structural social change. ("Subject" 53)

Berlant critiques this strategy on the basis that it problematically reaffirms institutional sites such as the law or nation-state, but we might additionally consider how the effectiveness of a politics premised on the possibility of a collective union based on shared affective "feeling" may itself be becoming moribund in a moment in which such sites and institutions are on the wane in general within a global economic system which often takes intense feelings and experiences as its fundamental "products."

Brian Massumi, despite his more optimistic readings of affect discussed earlier in this text, has also thematized this movement in direct relation to the decline of disciplinary techniques of power and problem of identifying techniques of resistance:

> It's no longer disciplinary institutional power that defines everything, it's capitalism's power to produce variety . . . The oddest of affective tendencies are okay—as long as they pay . . . [Capitalism] hijacks affect in order to intensify profit potential. It literally valorizes affect. The capitalist logic of surplus-value production starts to take over the relational field that is also the domain of political ecology, the ethical field of resistance to identity and predictable paths. It's very troubling and confusing, because it seems to me that there's been a certain kind of convergence between the dynamics of capitalist power and the dynamic of resistance. ("Navigating" 24)

As Massumi suggests, strategies based on forming collectives around affective intensities or shared feeling appear to be no longer the domain of resistance to social power; rather they seem to define their very logic of operation.

Third and finally, all of this is of a piece with the vast progression of communication and exchange networks, the organizing non-space of economic production and cultural representation. As Manuel Castells writes,

economic and social processes increasingly occur not in concrete locations, what he calls "the space of places," but the more flexible "space
of flows," the intersection of "purposeful, repetitive, programmable
sequences of exchange and interaction between physically disjointed positions" that structure contemporary "network" society (412).[3] In many
ways founding cyberneticist Norbert Wiener was perhaps the first to draw
our attention to this shift. As underscored in our first epigraph above,
Wiener hypothesized that the late twentieth century and beyond would
be defined not so much by particular physical machines and their outputs, but rather by the discrete capacities and forces of "communication
and control" foregrounded in cybernetic explorations of both machinic
and organic entities. In distinction to the automata of other centuries
that were defined best by their outputs, contemporary automata are constructed by maintaining flexible interconnections between human and
mechanical systems. As Wiener writes, if there is a representative machine
for the present age, it is the servomechanism, which allows one to exercise
control "at a distance" and which mimics physiological process of feedback and equilibrium that take place in physiological bodies (*Cybernetics*
43). As numerous social and cultural theorists have since foregrounded,
the domination of network logics has challenged not only our traditional
notions of economic production, but also our common conception that
forces of subjugation and subjectivity formation primarily take place
within sites of confinement or around particular identity categories. As
economic reliance on discrete products and services breakdown, institutional sites of disciplinary activity are also outsourced and distributed:
marketplaces, factories, prisons, and schools give way to e-commerce,
telecommuting, house arrest, and distance education.

Such a network logic helps explain the relatively novel constellations of
marketing and profitability emerging through such a dislocation of spatial
dimensions. To return one last time to the examples with which we began
this chapter, we might consider that the impoverished and the homeless in
particular have always occupied a rather problematic place in systems and
theories of disciplinary power. As Zygmunt Bauman suggests, the poor are
often conceived as the only individuals who fall outside of a seemingly axiomatic strain of capitalism, the non-producers/non-consumers that, on the
one hand, give lie to celebrations of the "mobile workforce" or "nomadic
subject" portrayed within analyses of capitalist labor and consumption,
and, on the other, serve mainly as motivation for others to be "good"
members of a capitalistic system (222).[4] It would take the faster and more
mutable technologies that develop in the wake of the breakdown of the sites
of disciplinary power—and operate through communication and control
practices in contrast to discipline's reliance on the body—to develop ways
to directly profit from the homeless and homelessness in such manifestations as the homeless vacation, homeless week, secret camera photos of the
homeless, and Bumfights.com.[5]

Precisely for these reasons, the networked nature of contemporary political economy has made it increasingly difficult to theorize how one might productively respond to structures of power based on the "spaces of flow" and flexibly mutating strategies of control, given that our typical hermeneutic and analytical techniques are based on the targeting of institutional bases and specific forms of subjugation. If the fundamental dilemmas of early postmodernism were based on the breakdown of binaries between particular categories (Marxian base/superstructure in economics, high/low culture in aesthetics, and knowledge/narrative in epistemology), the contemporary dilemmas of what I have been calling the cybernetic age appear to be instead clustered around the binary logic found in computational technologies, where elements are continually transformed and linked in flexible ways.

In the remainder of this chapter, I would like to explore possible responses to these problems in reference to the writings of two figures who were perhaps the most prescient in anticipating these changes and the corollary question of "resistance": Gilles Deleuze, who in two short but influential pieces suggested that "control societies" were taking over from Foucault's disciplinary models, and the American writer William S. Burroughs, from whom Deleuze borrows the concept of "control" as a sociotechnical system.[6] Although I work through the similarities in their thinking, I am ultimately interested in what we might learn from their divergences, and specifically what Burroughs might have to teach us in regard to rethinking categories of "resistance" within contemporary "control systems" of social power. In this latter endeavor, my objective is not to suggest that Deleuze somehow did not read Burroughs "rightly," but rather to follow what I take to be the Deleuzian line and see if this interaction can help us restate or reformulate the problems of control and resistance for the present.[7]

DELEUZE AND BURROUGHS ON DISCIPLINE AND CONTROL

In his book on Foucault as well as several of his final essays, Deleuze would draw on Foucault's own interest in the increasing obsolescence of disciplinary systems and borrow Williams S. Burroughs' concept of "control systems" to assay the "widespread progressive introduction of a new system of domination" he would alternately refer to "control," "communication," or "cybernetic" societies. Although Deleuze makes no explicit reference to particular works by Burroughs in these writings, there a number of ways in which Deleuze's work on post-disciplinary social power overlaps with Burroughs similar considerations about the nature of "control" in the late twentieth century.

Perhaps most obviously, both foreground a logic of "intensity" in charting changes in the dominant structures of social control—the tendency for existing mechanisms to trend toward greater efficiency by producing the same results achieved by earlier forms of power but through less brutal or more "cost-effective" methods. Deleuze approaches this tendency in reference to Foucault's earlier portrayal of disciplinary societies as itself something like an intensification of previous systems of power; as Deleuze writes, although Foucault recognized the unparalleled efficiency of discipline, he also recognized "the transience of this model: it succeeded that of the *societies of sovereignty*, the goal and functions of which were something quite different (to tax rather than to organize production, to rule on death rather than to administer life)" ("Postscript" 177). While discipline worked bodies in relation to physical sites such as the factory or school, control societies emerge as forces of social control become increasingly mutable and can operate at greater distances through control and communication technologies. In other words, the forces at play in disciplinary power have not so much been replaced as they have been perfected—made more efficient and flexible. As Deleuze writes, discipline worked by casting individuals inside of particular "molds" that form a body's identity and potentiality, but control works more as a generic type of "modulation" that varies from one moment to the next (178–179). As Michael Hardt suggests, the important distinction here is between the disciplinary construction of a "fixed social identity" and a new form of control that operates on a body "as a whatever identity": "Mobility, speed, and flexibility are the qualities that characterize this separate plane of rule. The infinitely programmable machine, the ideal of cybernetics, gives us at least an approximation of the diagram of the new paradigm of rule" ("Withering" 36). While the general breakdown of disciplinary sites initially seem to present new freedoms, they in fact lead to what Deleuze calls "mechanisms of control as rigorous as the harshest confinement" but with much greater flexibility and without the need for instances of actual physical confinement ("Postscript" 178).

Burroughs uses the term "control" exclusively in his analyses of social power, but likewise contrasts its current manifestations in contrast to harsher, more physical mechanisms of domination and persuasion. In a variety of his fictions these forces are embodied in particular characters, most notably a police team identified as Hauser and O'Brien in *Naked Lunch* but who circulate under the designations "the tough cop" and "the con cop" elsewhere in Burroughs' fiction and nonfiction. In Burroughs' narratives, Hauser is the tough cop, the one who immediately begins with intimidation or delivers a few punches to soften up the suspect. O'Brien, however, represents a method all the more effective for being less blatant, less prescriptive, and less painful. Yet, the two are indeed a "team"; lurking in the background of the con cop's pleasantries is the threat of violence implied by the tough cop. As Burroughs writes in "Academy 23," "the con cop's arm around your shoulder, his soft persuasive voice in your ear, are indeed sweet nothings without the tough cop's blackjack" (91). However, the con cop's willingness to make a deal allows the opportunity to negotiate your way out of the violence of the tough cop, and even to escape both cops altogether, even if only temporarily. What Burroughs suggests fictionally in this routine is made more explicit in the essay "The Limits of Control" where he emphasizes that most contemporary forms of social control operate through communicative practices that make control systems more flexible but also more open to manipulation. For Burroughs, "all control systems try to make control as tight as possible" but, "if they succeed completely, there would be nothing left to control" (117). If a control system eliminates any chance of rebellion, it also necessarily eliminates all agency in the subjects of control. For this reason, what Burroughs calls "control" must always operate through the use of communicative acts that function as commands:

> words are still the principal instruments of control. Suggestions are words. Persuasions are words. Orders are words. No control machine so far devised can operate without words, and any control machine which attempts to do so relying entirely on external force or entirely on physical control of the mind will soon encounter the basic limits of control. (117)

Control systems in Burroughs' schema share the basic logic of the networks they are carried through; like both biological and computational networks, they meet their highest efficiency at the edge of chaos.

For both Deleuze and Burroughs, such network schemas foreground another key difference between disciplinary sites of confinement and control systems: whereas in a panoptic schema some center, even an absent or implied one, must act as a point of surveillance or domination, control systems are based on an infinite series of linkages—the important point being not that control systems lack centers, but that they operate through multiple connections. Deleuze assays this difference through the array of connections that develop as disciplinary sites of confinement break down. In disciplinary arrangements, movement occurred between discrete sites: "Individuals are always going from one closed site to another, each with its own laws: first of all the family, then school ('you're not at home, you know'), then the barracks ('you're not at school you know'), then the factory, hospitals from time to time, maybe prison" ("Postscript" 117). As such sites decline, forces of social control become more diffuse and pervasive, increasingly "located" in transmissions of media of various types rather than within physical sites. Burroughs would organize such connections around the concept of "coordinate points," the locations where a force of control intersects a subject (*Nova* 57). As his collaborator Brion Gysin writes, much of Burroughs' work was premised on "disconnecting and reconnecting" such intersections (Burroughs and Gysin 17). Burroughs experimental writing techniques such as the cut-up and the fold-in were designed to destabilize well-worn associations between phenomena and then forge new connections by plugging "normally dissociated zones into the same sector." The specificity of such linkages as well as the multiplicity of their collective operations are crucial to both Burroughs and Deleuze insofar as they suggest a way to thematize social control as a distributed and continual process working through a network, as opposed to a discrete series of operations functioning in relation to particular sites and in the service of producing specific states of identify or subjectivity.

In the most basic terms, then, Deleuze and Burroughs help us to rethink the structure of social control as it appears in the wake of an ongoing breakdown of traditional institutional sites. However, substantial differences remain in the ways they will conceive their own style of engagement or intervention in relation to this shift, differences that are worth examining at some length insofar as they shed light on the possibilities for answering the question of "resistance" today or thinking through strategies for responding to diffuse systems of social power. If, as mentioned near the start of this chapter, the network logic composing forces of subjection has posed many challenges to traditional strategies developed in response to disciplinary power, my argument will be that examining this difference may give us a new angle on rethinking such techniques as well as the category of "resistance" itself.

CONTROL, THE COMMON, AND *DOXA*

As I argue above, Deleuze and Burroughs offer us a way to understand the changing shape of social control as a network system produced by the coeval development of capitalist modes of production and information technologies. Though the diagnosis of this situation is fairly similar in both accounts, there is an important divergence in the methods that each suggest for responding to such a system, one that I will argue might be thematized most productively around vectors of persuasion historically associated with *doxa*, that ancient rhetorical concept that often circulated as a counterpoint to *episteme* in ancient Greek culture (the "mere opinion" that was often portrayed as double or shadow of the "true knowledge" being produced by early philosophical thought and the emergent natural sciences). Indeed, it seems to me that *doxa*, while not having necessarily experienced the kind of renaissance that other early Greek rhetorical concepts have in recent cultural and political theory (such as *kairos* or *ethos*), is perhaps the missing term in a large number of recent discussions about the increasingly decentralized, "post-ideological," or post-disciplinary nature of contemporary social power.

First, it seems like the notion of *doxa*—as a general or surface contention shared across the social field—can do some heavy lifting around the dilemma posed by the last of the three shifts in social power as described by Foucault and Deleuze: from social power centered on the application or threat of direct physical force as managed by a central authority (sovereign power), to its application via a subject's training in particular epistemologies and performance of particular behaviors at multiple institutional sites (discipline), to its contemporary formation as the decentralized and feedback-based interactions between individuals and diverse forms of institutions, media, and technologies (control). *Doxa*, I take it, would be one of our best extant concepts to describe the peculiar structure of the last of these systems, one in which more superficial and action-oriented motivations and dispositions are managed, as opposed to in the more brutal regime of sovereign power and the conflation of identity with epistemologies in disciplinary regimes (as described by Foucault's famous "power/ knowledge" configuration).

In this regard, we might consider how *doxa*, insofar as it connotes shared and fluctuating contentions rather than more traditionally veridical standards of knowledge or belief, has long been an implicit component of our analyses of social power, even if playing only a largely negative or ancillary role. In the Platonic critique of *doxa*, or at least in the version of it that was often invoked as a contrasting figure in poststructuralist thought, the problem with *doxa* is that it shows the negative influence of social factors on an individual's knowledge. Since *doxa* is by definition associated with the popular beliefs of "the crowd," it remains the corrupt version of, or constant threat to, true knowledge as understood and proposed by those within the relatively singular or elitist class of the philosopher. However, in the poststructuralist critique of the legacy of this formation, one in which the role of power in legitimating what counts as knowledge or truth, *doxa* takes on a somewhat more productive role. Or, perhaps more precisely, we might say that *doxa* becomes the negative force through which all epistemic structures—all of those formations that would claim to be true knowledge and thus delegitimize opposing structures as "just" *doxa*—are interrogated for the degree to which they themselves are merely forms of *doxa* that have risen to the category of knowledge at different historical moments and under different frameworks of legitimacy or veridicity. On the one hand, then, *doxa* has long provided the alternative category against which dominant forms of knowledge and belief can be compared. On the other, the association of *doxa* with "common" or popular opinion, often disguised as knowledge, lives on in a variety of efforts in twentieth-century critical and cultural theory that drew our attention to the ways in which socially sanctified or pervasive beliefs were both those that seemed to occupy the role of knowledge and *episteme*, but were *also those* that, precisely because of their ubiquity or "common sense" nature, should be the targets of critical powers of contestation and demystification. At least since the time of Adorno and Horkheimer's famous (re)definition of "Enlightenment" as "mass deception," as merely the moment in which a large enough group of individuals have been *deceived in the same way*, one of the shared contentions of subsequent critical theory—one is tempted to say, its own *doxa*— was of the necessary, and at least potentially revolutionary, task of showing the doxological structure of all dominant forms of knowledge.

Indeed, one could even expand this comparison to argue that the entire history of *both* philosophy as the study of constitution of knowledge *and* politics as the study of how power functions within social systems—even before the two become so explicitly entwined in mid-twentieth-century intellectual thought—shows them to be mediated by recurrent accusations of, and a fundamental reliance on, *doxa* as category. The historian of intellectual thought David R. Kelley has suggested something of the same in an essay that precedes through the brilliant conceit of visiting a number of topics associated with *doxa*—"mere opinion," "prejudice," "the role of history"—twice: once from the perspective of the mainstream, post-Platonic, anti-*doxa* strain of Western intellectual thought, and again in reference to a possible parallel "philodoxy" composed around thinkers who insisted on the necessity of opinion as a part of the accumulation and dissemination of knowledge. Through such a rereading, *doxa* emerges as a crucial category in both cases, whether as a necessary target for philosophy or as the unstable center of a "philodoxy" that still strives to measure the possibilities of consensus knowledge and its social effects. However, I take it that the question of *doxa* is different today than when considered as part of either the above "short view" of the role of *doxa* in ancient Greek culture and its return in poststructuralist thought, or even in any "long view" that would capture its consistent, if implicit, position in Western intellectual culture up to the late twentieth century. Both of these cases tend to emphasize the importance of knowledge or some kind of ideological formation of presupposed beliefs as central, even if they are the same ones waiting to be revealed as merely *doxa* that was created or sustained by the particular structures of social power of a given time. The challenge in the present case is perhaps to think through the consequences that occur when *doxa* becomes *the central* category of contemporary sociality and public persuasion.

Second, *doxa* would seem to be a concept tailor-made for the otherwise difficult to describe ways in which contemporary social power works through a series of dynamic and feedback-based connections between individuals and broader social forces. Perhaps the most challenging (and therefore controversial) component of Foucault's theorization of social power around the concept of discipline was the way in which it multiplied the sources and methods through which the framework of commonplace beliefs or values circulated within societies. For one, it did not offer the relative comfort, at least from the point of view of the critical theorist or activist who might wish to operate starting from this conception, provided by a more traditional notion of ideology as a ruling concept; as opposed to earlier forms of ideology critique, one could not posit that an individual is conscious of only a distorted view of their material reality, and one that might be recovered in some way by retracing the path through which ideational frameworks of a particular culture emerge from and cover up their material conditions.[8] Nor, for that matter, and despite the associations often made between Foucault's thought and "social constructionist" perspectives in cultural theory popular around the same time, does disciplinary power seem to provide us any of the benefits of that view of how social power is formed and maintained. In identifying social power as an inherently positive force structuring social relations the contemporary forms of subjectivity, Foucault cuts off any easy conclusion that such categories are merely the result of a kind or reasoned consensus of contemporary individuals (which we might "reason" against or rationally think our ways out of) or simply the dead weight of history deploying its residual effects on the present (in which case we could posthumously condemn past actors for creating the negative effects of social power that we experience in the present).

Although it was this kind of highly complex and capillary version of social power that was the major target of critiques of Foucault's mid-career work—the realization that power is felt as a pervasive quotidian experience bundled with the contention that this realization itself will in no way negate its functioning and effects—we might say that the updating of social power as "control" in the works of Burroughs and Deleuze is, if anything, *even more capillary* and *less suggestive* of effective strategies of resistance or redress. Even if disciplinary power's focus on the construction of varied, but particular, subjectivities made it appear immune to older forms of analysis and critique, it still suggested at least one option: working through and against such forms of identity as a way to expand the boundaries of, or multiply the possibilities within, what seemed like fairly rigid and endlessly reaffirmed options for individuals within disciplinary formations.[9] In what Deleuze and Burroughs refer to as societies organized around "control," however, this kind of flexibility is already an inherent part of the system; indeed, such work of expansion and multiplication seems to be in many ways what, for lack of a better word, social power "wants" from individuals, rather than the best way to disrupt its smooth functioning. Simply put, since in control societies, individuals are not merely "made subject to" forms of power, but in many ways provide the content and direction of more generic forces arranged around discrete activities or outputs, there does not seem to be much purchase on attempting to shore up either the "objective" or "subjective" nature or status of any particular ideology, discourse, or thematic of any type, as a way to get purchase on an outside or oppositional perspective or practice on control.

And it is in precisely this kind of situation that a focus on a kind of "opinion" as the modulation of common and surface-level shared affectations named by *doxa* can do a large amount of work around figuring out both the nature of power and "resistance" in the present. In particular, *doxa* has long designated the kind of co-constitutional domains of subjective/objective or individual/social experiences and motivations in a way that is difficult to thematize in recent theories of power. The classicist Eric Havelock traces this history from its point of origin:

Both the noun [*doxa*] and the verb *doko*, are truly baffling to modern logic In their coverage of both the subjective and objective relationship. The verb denotes both the "seeming" that going on in myself, the "subject," namely my "personal impressions," and the "seeming" that links me as an "object" to other people looking at me—the "impression" I make on them. The noun correspondingly is both the "impression" that may be in my mind and the "impression" held by others of me. It would appear therefore to be the ideal term to describe that fusion or confusion of the subject with the object. (250–251)

As I have been arguing throughout this chapter as well as various earlier points in this book, a very large share of phenomena shaping contemporary forms of social power—from media networks, to niche marketing, to populist discourse—seem to function through this kind of feedback-driven blurring of such categories and the particular forms of attraction and influence, ones best captured by persuasion as a motivating force and rhetoric as a disciplinary domain, that seem inherent to such an arrangement.

In the remainder of this chapter, I want to take up such a *doxa*-centered approach in thinking through the second segment of our itinerary here, thematizing the possible presentations of "resistance" suggested by Deleuze's and Burroughs' respective thought as the counterpoint to the description of power described above. To provide a bit narrower of a focus, I will be concentrating on marketing as a particularly exemplary phenomena of control societies. In a broad sense, this might seem like a fairly obvious focus; much as advertising often played the role, in the work of such thinkers from Roland Barthes to Raymond Williams, of an earlier mode of power based on making associations between its audiences and socially pervasive categories of desired status or behavior, marketing, that more feedback-driven and aleatory form of promotion via media, is perhaps most representative of the logic of control in its most quotidian and pervasive forms.[10] More specifically, as I will suggest below, marketing is perhaps the best point of entry into determining the contrast between Deleuze and Burroughs on the question of "resistance" in control societies, one that I will further argue might be particularly important for modeling our own responses or interventions around control in the present.

A BRIEF COMMERCIAL INTERRUPTION

We might begin unpacking Deleuze's engagement with these vectors around his strategies for what would constitute "liberation" or "resistance" to both the negative effects of capitalism and contemporary social control, strategies that have themselves shifted through a number of works. If, as discussed above, contemporary discussions about the definition, or possibility, of "resistance" in contemporary cultural theory often end up invoking in one way or another the late works of Foucault—and, more specifically, the question of whether his late works seem to suggest a certain complicity with the forces of power that were the targets of earlier texts—something of the same critique has appeared in reference to Deleuze's writings on politics and social power. Slajov Zizek has provided perhaps the most pointed version of this critique; referencing a book review of Deleuze and Guattari's *What Is Philosophy?* that imagines the puzzled look appearing on the face of "enlightenment-seeking yuppie" that has purchased the bestselling book expecting a pop-philosophy text and is now confronted with "page after page of vintage Deleuze," Zizek imagines the opposite reaction:

> What, however, if there is no puzzled look but enthusiasm, when the yuppie reads about impersonal imitation of affects, about the communication of affective intensities beneath the level of meaning ("Yes, this is how I design my publicities!"), or when he reads about exploding the limits of self-contained subjectivity and directly coupling man to machine ("This reminds me of my son's favorite toy, the action-man that can turn into a car!"), or about the need to reinvent oneself permanently, opening oneself up to a multitude of desires the push us to the limit ("Is this not the aim of the virtual sex video game I am working on now? It is no longer a question of reproducing sexual bodily contact but of exploding the confines of established reality and imagining new, unheard-of intensive modes of sexual pleasures!"). (*Organs* 183)

Such enthusiasm, Zizek argues, suggests that a variety of the critical modes and practices Deleuze had developed as methods of resistance to the negative effects of capitalist production are in fact "features that justify calling Deleuze the ideologist of late capitalism" (183–184). Such a contention is an overstatement, of course, particularly insofar as it ignores Deleuze's well-known disaffection for both ideology as a critical concept and his rather complex engagement with capitalism as a system (of which, more in a moment); however, I am not so much interested in this particular reading as I am in exactly what might constitute methods of "complicity" or "resistance" in reference to the shifting grounds of contemporary capitalism and social control, methods Deleuze has reformulated several times in the space between the volumes of *Capitalism and Schizophrenia* and his later writings on control systems, and one that might be read as microcosm for the problematics of rethinking "resistance" in the contemporary moment.

As numerous critics have mentioned, the years in between *Anti-Oedipus* and *A Thousand Plateaus* seem to indicate not only stylistic but also strategic changes in Deleuze and Guattari's thinking. The first volume progressed toward a "schizoanalytic" methodology that would seek to liberate desire by consistently deterritorializing the social coding of capitalist subjection; though capitalism is depicted as itself an immense system for releasing unpredictable forms of desire, it continually reterritorializes in the creation of increasingly exploitive systems. However, the schizoanalytic method of consistent deterritorialization risks being outpaced, as in the interventional strategies mentioned above, as it appears geared toward an understanding of social control that produces discrete normative states rather than the flexible modulations of contemporary capitalism. As Eugene W. Holland argues, Deleuze and Guattari's own attention to this change appears to emerge between the volumes and helps explain the absence of schizoanalysis in *A Thousand Plateaus*. As Holland writes, "as long as the relative fixity of social codes—codes of labor discipline, codes of collective fashion preference, and so on—was crucial to axioms of mass production, schizophrenic decoding had a point (*a point d'appui*) and a disruptive effect," but as codes of social control became more modular rather than modal, a new strategy was needed to rethink how it might be possible to reorganize lines of flight and modes of resistance in more productive ways and in more specific circumstances (72).

Indeed, when Deleuze later begins diagnosing new forms of capitalism and the control society model of power explicitly, the enemy shifts from the metageneric process of rigid reterritorialization to a very specific technique: marketing. Though Deleuze leaves the question of what role opposition may take in control societies, he concludes "Postscript" by stating that "future forms of resistance" will be those "capable of standing up to marketing's blandishments" (182). Earlier, he positions marketing as the key vector of contemporary control and the source of its negative effects: "Corruption here takes on a new power. The sales department becomes a business' center or 'soul.' We're told business have souls which is surely the most terrifying news in the world. Marketing is now the instrument of social control and produces the arrogant breed who are our masters" (181). This focus on marketing as primary content provider for social control of course makes a certain amount of sense; if flexible control and finance capitalism operate not through the imposition of discrete normalizing states and modes of subjection, but rather through a consistently varying response to potential new markets and supple subjectivities, then it is certainly reasonable to foreground marketing as the primary vector for these control systems. However, my concern here is with the degree to which one should, or can, hypostasize marketing as a *technique* or mode of engagement that must be resisted.

Deleuze and Guattari thematize marketing in much the same way throughout *What Is Philosphy?*, published around the same time as Deleuze's writings on control. Here marketing is positioned precisely as an opposition to the productive force of philosophy:

> Finally, the most shameful moment [for philosophy] came when computer science, marketing, design, and advertising, all the disciplines of communication, seized hold of the word *concept* itself and said: "This is our concern we are the creative ones; we are the *ideas* men! We are the friends of the concept, we put it in our computers" (10)

It, of course, makes sense to *distinguish* between philosophy and marketing in a primer that is primarily geared to define philosophy for a general audience. However, it is again the *oppositional* nature of this distinction that I want to question. As Deleuze and Guattari formulate it, philosophy "has not remained unaffected by the general movement that replaced Critique with sales promotion" but the crowding out of philosophy is a call for the reinstallation of philosophical critique: "Certainly, it is painful to learn that *Concept* indicates a society of information services and engineering. But the more philosophy comes up against shameless and inane rivals and encounters them at its very core, the more it feels driven to fulfill the task of creating concepts that are aerolites rather than commercial products" (11).

Deleuze and Guattari's disaffection for marketing is related to their treatment of opinion in *What Is Philosophy?*, a category that is similarly the target of much acrimony. As Deleuze and Guattari detail, the philosopher is confronted from all sides by the forming and circulation of opinions; the challenge for the philosopher is to extract "knowledge" from them without either submitting opinions to a higher standard of valuation such as Platonic or Hegelian dialectic, which, they write, leads only to "interminable discussion," or applying "universals of communication" such as Aristotle's topics to the study of them (79). The problem with opinions, they write, is that they formulate thought around processes of "recognition" that seem to divide individuals into groups based on commonality or rivalry. In its worst manifestations, working with opinions can lead to the creation of monolithic or homogenous collectives through the establishment of orthodoxy. However, this "is still only the fist step of opinion's reigns: opinion triumphs when the quality chosen ceases to be the condition of a group's constitution but is now only the image or 'badge' of the constituted group . . . Then marketing appears as the concept itself" (145). For Deleuze, then, it seems marketing is not only the engine of capitalism and the locus of its negative effects, but also the technique that must be distinguished and resisted, particularly insofar as it continually brings us into engagements structured largely on the grounds of opinion.

I dwell on Deleuze's treatment of marketing above because I believe it helps underscore the contemporary problematics of articulating "resistance" in response to control societies. If, as argued at various points in this book, traditional critical strategies—interventions based on foregrounding difference, situating knowledge claims within their ideological and social contexts, and exploring the possibilities of leveraging affective forces against capitalist exploitation—appear to lose their effectivity in response to the increased flexibility of parametric modes of social power, then "resistance" seems to be stuck being positioned within some logic outside of this ostensibly all-consuming flexibility, or it can only find its identity in being the force that blocks the persuasive power of such a system.

I find such a move troubling insofar as it seems to suggest that as flows of social control become *increasingly flexibly*, modes of resistance must become themselves more *flexibly oppositional*. However, as social control becomes increasingly networked and less reliant on rigid sites and discrete states, there becomes literally nothing or at least "no one thing" to resist, which suggest we may have to rethink the category of "resistance" altogether. We might begin such an itinerary by returning again to the domain that seems to largely factored out or made abject in Deleuze's analysis: rhetoric. It seems to me, as suggested by Deleuze and Guattari's disaffection for engagements with opinion, their strategy equally opposes the application of rhetoric as an ecology of strategies and techniques. Most generically, it seems to undermine the force of rhetoric in its productive rather than interpretive form, as, in its canonical definition, to identify and mobilize all possible means of persuasion in a given situation. More specifically, but in a similar manner, it equally discounts the force of rhetoric as a strategy that enacts persuasion by working *through* the established opinions of a given audience or the given forces circulating in an environment. Of course, if as Deleuze and Guattari suggests, the "best" elements of philosophy have been appropriated by marketing, this is certainly at least as true of its cooption of rhetorical techniques.[11] I take it, however, the fundamental rhetorical move would be to develop methods for working through the logic of marketing insofar as it seems to be the dominant force and ecology of the persuasion in the present moment.

This is, I think, where Burroughs' intervention is crucial. As detailed above, Deleuze and Burroughs present largely similar diagnoses of social control; however, Burroughs' prescriptions would be consistently organized around strategies of inversion or reversal that attempt to work through the mechanisms of control to produce certain effects. As he writes in "The Revised Boy Scout Manual," "the extent to which revolutionary theory and tactics is disadvantageously shaped by *opposition* is something few revolutionaries like to think about," and we might take up Burroughs' contrary appropriation of control mechanisms in a variety of registers as inspiration here (5; emphasis added). For instance, stylistically, though Burroughs' early writings made extensive use of cut-ups, random assemblages of words and texts that would block typical associational patterns of reading, he would later express "complete dissatisfaction" with this work and its tendency to leave him in "complete isolation" away from dominant narrative styles (qtd. in Ansen 55). Instead he would create strategies for producing the same effects through more common modes of writing, for instance, not only using "fragments of [Joseph] Conrad" in cut-ups but also adapting "the Conrad Style" (qtd. in Harris, 259). In the control economies of his fictions, he would introduce Clem Snide, Private Asshole, who uses the methods of the "tough cop" and "con cop" for his own purposes. Appropriately, when assigned to rescue an abducted client, he seeks demographic research: "I want to know preferences in food, clothes, colors, reading, entertainment, use of drugs and alcohol, what cigarette brand he smokes, medical history" (*Cities* 37).

Again, though marketing is taken up here only as an example of this strategy of working through control, it is perhaps a particularly important one given the concerns of this chapter. We might begin unpacking Burroughs' relation to marketing tactics in reference to Thomas Frank's response to Burroughs' 1995 appearance in a Nike commercial, one that is at least partially comparable to Zizek's critique of Deleuze's complicity with capitalist modalities. In "Why Johnny Can't Dissent," Frank similarly argues that Burroughs' commercial appearance is an indicator of larger changes in contemporary capitalism:

> The most startling revelation to emerge from the Burroughs/Nike partnership is not that corporate America has overwhelmed its cultural foes or that Burroughs can somehow be "subversive" through it all, but the complete lack of dissonance between the two sides. Of course Burroughs is not "subversive," but neither has he "sold out": His ravings are no longer appreciably different from the official folklore of American capitalism. What's changed is not Burroughs, but business itself. As expertly as Burroughs once bayoneted American proprieties, as stridently as he once proclaimed himself beyond the laws of man and God, he is today a respected ideologue of the Information Age, occupying roughly the position in the pantheon of corporate-cultural thought once reserved strictly for Notre Dame football coaches and positive-thinking Methodist ministers. His inspirational writings are boardroom favorites, his dark nihilistic burpings the happy homilies of the new corporate faith. (36–37)

Frank goes on to compare Burroughs' work to management guru Tom Peter's 1987 primer *Thriving on Chaos*, which he takes to anticipate the flexible logic of contemporary capitalism. Although Frank may be exaggerating Burroughs' importance to corporate culture—it is hard to imagine a circle of business man pounding thousand dollar watches off the board table chanting "You thing rike jellyfish pretty soon now" or "slow masturbation used to be me mister"—his comments illustrate an important point and one that I take to be somewhat different than Zizek's critique of Deleuze, or at least one that casts this kind of finger-pointing criticism in general in a different light. Whereas Zizek may be at least partially correct in stating that Deleuze's strategies of resistance are counterproductive insofar as they become, as compared against Deleuze's own analysis, similar to strategies of capitalist production, it seems to me that Burroughs' similarity in the same respects is precisely what makes his work productive. This is not to say that Burroughs was simply a canny manipulator of ways to generate profit, though—from his various chinchilla-, corn-, and marijuana-growing schemes early in his career (detailed in *Junky* and fictionalized in several of Kerouac's books), until the time his writing and celebrity became profitable—Burroughs was certainly never opposed to making a quick buck. More importantly, marketing would be one of many structures he would appropriate as a means of navigating the "control mechanisms" of contemporary media and dominant forms of social persuasion. For Burroughs, the problem was never these systems themselves, but rather finding a way manipulate the logic by which they work.

Appropriately, Burroughs' first paid position was in marketing. By his own account, after completing graduate work in anthropology and finding academic life to involve "too much faculty intrigue" Burroughs worked as copywriting in an advertising agency (qtd. in Knickerbocker 76). As he states later, he has "thought a great deal about advertising. After all, they're doing the same sort of thing. They are concerned with the precise manipulation of word and image"; but Burroughs is not interested in getting readers to "go out and buy Coca-Cola," but in creating "an alteration in the reader's consciousness" (81). In the same interview he argues that there is "no reason why the artistic world can't absolutely merge with Madison Avenue" (69). Again, Burroughs' interest in marketing tactics is one particular example of Burroughs' strategy for hacking control, but it is perhaps the most prescient one if marketing is becoming both the dominant and most flexible instrument of social control. The crucial question, as I have been arguing throughout this chapter, is not so much how these forces can be resisted, but to develop the possibilities for their manipulation. As Burroughs will suggest throughout his work, as counterintuitive as it may initially seem, the fact that the forces of social power have become increasingly ingrained into a vast variety of quotidian techniques that pervade contemporary culture, is also precisely what makes modes of effectively responding to them increasingly available and apparent. Or, as Burroughs writes in *Nova Express*, "The counter move is very simple—This is machine strategy and the machine can be redirected" (74).

OPERATION REWRITE: RHETORICAL
STRATEGIES FOR CONTROL SOCIETIES

All of the above, of course, leads us with many unanswered questions regarding how such a strategy of inversion or reversal might be mobilized. Though I take it this is perhaps *the question* of thinking through the complexities of social power in the present and one that is unlikely to be resolved all that quickly, in an homage to Burroughs' penchant for the programmatic register—"*Naked Lunch* is a blueprint, a How-to Book" (182)—I offer below three possible instructions for working through forces of social control.

Who Am I to Be Critical? Critical Theory
after "the Subject" of Power

Certainly one of the more difficult dilemmas posed by both Deleuze and Burroughs in their survey of control societies is the ways in which contemporary social power seems to feed on the very tools of critique—the challenging of essentialist binaries, the exposure of the bias or contradictions within dominant modes of thought and belief, the multiplication of categories of acceptable identity or behavior—that drove much politically oriented cultural theory of the previous decades. Here too we might read this dilemma as the repetition, at another level, of the same problems posed by Foucault's disciplinary power; one of the major critiques of Foucault's theorization of social power was that its capillary conception of power as a force that *made subjects* by *working on bodies* seemed to leave very little room for any kind of political intervention that relied on somehow "returning" people to some kind of originary or natural identity existing underneath the constructions of social power, or, as in the identity politics popular at the time of Foucault's writing, trust that critical analyses of the various ways that particular social groups have been disadvantageously defined by such constructions could pave the way toward more egalitarian forms of representation and community. Though, as mentioned above, the shift from analyzing the category of (ideological) identity to (social) subjectivity and a more free-floating and diverse style of critique in general—one in which communication and representation "themselves" rather than more rigidly defined systems of economic or imperialist exploitation became targets—was salutary, even these avenues of intervention seem to be cut off in the transition from discipline to control. In a time wherein the criticism of traditional scientific rationality is more likely to be voiced by conservative political strategists than it is by left-oriented humanities and social science scholars, and while moving through your radio dial you are just as likely to hear about the latest governmental or social oppression of wealthy white American males as you are of any other group, both critique as a method of epistemological exposure and "the subject" as a focus for tracking social power seem to have lost much of their effect.

For these reasons, perhaps one way to begin responding to these shifts is to reimagine the role of critical and cultural theory as performing a function and a focus other than that of these well-worn vectors. Perhaps it is not even so much that critique has been hijacked by the factions it was typically wielded against in critical theory that is the problem here, inasmuch as it is critique as a mode that presumes to show the mistaken epistemological assumptions of a particular group has always been a pretty fraught undertaking, becoming in large part, as Jeff Nealon suggests, merely "a version of the condescending convention wisdom that most people don't know what they're doing" (*Foucault* 108). This is not to say that "critique" of this type does not do a large amount of work—that is precisely why we might presume it has been appropriated by such diverse parties—but perhaps the real problem here is coming to terms with how even it has always functioned as a motivational force, a form or persuasion rather than of enlightenment.

Concomitantly, we might do well to spend more time thinking about how it works on that kind of surface level, as a way of directing and refining certain existing dispositions and common conceptions already extant in particular social fields. This is, I take it, the lesson we might read in Burroughs' nonfiction writings on social control as well as the shift in his aesthetic work from experimental writing based on the appropriation and rearrangement of particular words and images to his appropriation of whole "styles" of expression and persuasion. This too, I take it, was an important focus of an earlier vintage of Deleuze. Consider, for instance, Deleuze and Guattari's description, in *A Thousand Plateaus*, of how to experiment with existing forces of social power by "making yourself a Body without Organs (BwO)":

> This is how it should be done: Lodge yourself on a stratum, experiment with the opportunities it offers, find an advantageous place on it, find potential movements of deterritorialization, possible lines of flight . . . It is through a meticulous relation with the strata that one succeeds in freeing lines of flight, causing conjugated flows to pass and escape and bringing forth continuous intensities for a BwO. Connect, subjugate, continue: a whole "diagram," as opposed to still signifying and subjective programs. (161)

What we might learn from both Deleuze and Burroughs, then, is to "diagram" such forces *as forces*, as the effects of particular techniques and strategies, rather than the creation or destruction of epistemological frames or of the possible identities and subjectivities of individuals.

The Long Con; or, the Scrivener and the Confidence Man

We might draw a related strategy from Burroughs' work in and around the ways in which he tends to present techniques for intervening in contemporary control systems as themselves largely appropriations of the dominant logic of that system—a kind or "reappropriation" or act of reverse-expropriation performed in response to how he sees control systems as themselves acting parasitically on human capacities for imagination and invention. In Burroughs' work, such subversive or inversive acts are typically performed through various "cons" or confidence games; the techniques of social power that Burroughs personifies in the "con cop" character that appears in his fiction is frequently met by a number of "con men" and related figures whose deploy the same techniques in pursuit of different or opposite ends. Though we have seen some examples of the above in Burroughs' work already, in regard to relating this technique to the larger question of "domination" and "resistance" in social power today, we might shift to two characters from the fiction of another author much beloved by both Burroughs and Deleuze, one of which has already taken on a certain symbolic value as a symbol of resistance in contemporary cultural and political theory.

I refer specifically here to the rather strange revival of the titular character of Melville's "Bartleby the Scrivener: A Story of Wall Street" as a heroic figure in writings of such thinkers as Agamben, Blanchot, Derrida, Hardt and Negri, and Zizek.[12] In all these appropriations, Bartleby's famously cryptic deferral—"I would prefer not to"—is taken as an example of absolute refusal or pure negativity, either as the clearing point for a new social arrangement to come, as some type of escape from the reciprocity of domination and resistance, or as a refusal to play the game of capital or of power. Bartleby's "absolute refusal" becomes the only authentic act available to an individual when every available action seems automatically compromised by its complicity with ever more flexible systems of social and economic interpellation or expropriation.

However, we might find a better or more appropriate figure for the present in another of Melville's characters, the polymorphic creation of his final novel *The Confidence-Man*. If "Bartleby" was a tale of how charity fails in an instance—the narrator relates how his charitable actions toward Bartleby are unsuccessful—*The Confidence-Man* indexes the impossibility of wholly "resistant" or "authentic" action in a time wherein all behaviors seem to already be complicit within the dominant power structures of their own environment. Through the series of vignettes that compose the novel, charitable actions are mixed with cruelty or revealed to emanate from necessarily ulterior motives: a proto-socialist "man in gray" launches trenchant critiques of capitalism while simultaneously dreaming of a "benevolence tax" that would be dedicated to easing poverty (later seeking donations for an orphan asylum that does not exist); a panhandler is ignored by a crowd clustered around the wanted poster of a criminal poster and buying lurid crime pamphlets; a disabled "New Guinea beggar" is only able to solicit charitable donations by degrading himself by catching thrown coins in his mouth and then is almost seized and beaten by a crowd of travelers after another panhandler alleges that his disability is feigned, and that he may be a white man in disguise. Anytime any charitable action takes place, it is revealed to be cathected to some notion of profit on behalf of the giver: a public display of their moral values, the storing of credit in heaven, or the chance to feel better about oneself as an ethical individual.

Though written long before the various changes in technoscience and political economy that have been the focus of this book, and most often approached through the various ways it reflects tensions (over identity, race, and social justice) of his own time, Melville's work in this text may be one of our best depictions of the challenges posed by the overwhelming *technicity* of the present: one in which persuasive techniques and strategies of all types seem to have become unmoored from their ideological backgrounds and made available for purposes quite contrary from their original purpose, one in which authenticity and acts of critique or judgment can obtain value only within the same economy of flexible power and persuasion they were once positioned against.

Any Number Can Play: The Technics of Power

Taken together, these strategies at best offer not "resistance" to the powers of control and communication, but rather a certain toolbox for manipulating these flows, an application of them through a logic of inversion and appropriation. We might borrow Deleuze's analogy for the modulation of social power— "*Surfing* has taken over from all the old *sports*"— and develop practices not in the direction of resistance but toward the creative "surfing" of these flows ("Postscript" 180). Here again, it is important to recall from the introduction of this book the coining of cybernetics as a discipline from the Greek term for "steering" or "navigation." Insofar as cybernetics seems to occupy both the form and function of control societies, the flexible strategies of feedback might be our best conceptual resource in rethinking what it would mean to "resist" such systems. Additionally we might emphasize the coincidence of this term with rhetoric in Wiener's source, Plato's *Gorgias*. The strategies of both cybernetics and rhetoric, it seems to me, imply not a resistance to or renunciation of the forces in which one is immersed, but instead techniques for their practiced redirection and the opening of occasions to respond to them differently. It is at this vector, I'd like to suggest, that we must begin searching for our ethical and rhetorical responses to the networked logic of control societies and of life in the cybernetic age. As Burroughs reminds us at the end of *The Ticket That Exploded*, the forces of control are open to "any number" of manipulations and appropriations: "The techniques and experiments described here have been used and are being used by agencies official and non official without your awareness and very much to your disadvantage . . . any number can play" (215).

5 On the Genealogy of Mortals; or, Commodifying Ethics

> Fixing prices, setting values, working out equivalents, exchanging—
> this preoccupied man's first thoughts to such a degree that in a cer-
> tain sense it *constitutes* thought . . . Perhaps our word 'man' (*manas*)
> expresses something of this *first* sensation of self-confidence: man
> designated himself as the being who measures values, who values and
> measures, as the calculating animal as such.
>
> —Friedrich Nietzsche, *On the Genealogy of Morals* (78)

> You see, what I want to do is not the history of solutions—and that's
> the reason why I don't accept the word *alternative*. I would like to
> do the genealogy of problems, of *problématiques*. My point is not
> that everything is bad, but that everything is dangerous, which is not
> exactly the same as bad. If everything is dangerous, then we always
> have something to do. So my position leads not to apathy but to a
> hyper- and pessimistic activism.
>
> —Michel Foucault, "On the Genealogy of Ethics:
> An Overview of Work in Progress" (231–232)

Although the last several chapters have engaged a variety of ways in which
the social and political impact of contemporary technologies and new communication media have intersected questions of ethics, in this chapter I turn
to what we might call the question of ethics "itself": the degree to which
contemporary technics might force us to rethink the very concept of ethics,
whether we take it to be the critical consideration of "right" and "wrong"
behavior or as the more general mapping of the nature of relationships
between individuals within a community. In the following I argue for the
need for such a rethinking by interweaving two narratives: one that traces
the emergence of Western ethics in Greek antiquity as indexed against the
techne and media of that time and its continued legacy in the present, and
one that trace the recent "turn to ethics" in contemporary political theory
and its relation to the dominant media and technologies of today. These
two threads will crisscross at a variety of points, but perhaps most importantly around the ways that *technicity* has relentlessly, if often implicitly,
shaped our understanding of the "mortal" world of the material present,

and the "immortal" realm of the ethical imagination and our ability to imagine the ideal society, and thus our understanding of both the present as well as our hopes for the future.

NOSTALGIA FOR THE FUTURE

In the mid-1980s Fredric Jameson argued that one of the defining vectors of postmodern culture was a "nostalgia for the present," a kind of temporal and representational short-circuit through which attempts to represent contemporary social reality seemed to result only in "the recombination of various stereotypes of the past" (*Postmodernism* 296). For Jameson this condition was both an epistemological blind spot caused by an inability to "think historically" as well as a compensatory mechanism—a privileging of the pleasures of the present in the face of an uncertain future. Around two decades later, as many of the hallmarks of what we called postmodernism seem to be disappearing, we might say that our own contemporary appears marked by a different, but equally inverted, temporal relation: a "nostalgia for the future." The present no longer seems to cloud our access to the past as much as contemporary cultural and economic life have become haunted by fantasies of, and investments in, the future. Concomitantly, speculations of a time to come increasingly seem to crowd out concern for the here and now. In other words, our feelings of "homesickness" or yearnings for another time or place increasingly focus less on the "good old days" as they do on something like "the good days yet to be."

It is tempting to see such a condition as little more than the latest variation on a much longer tendency, the same way that we might read the nostalgia mode under review by Jameson as marking the moment when the fairly long-standing sentimental vintage of nostalgia as an idealization, and thus distortion, of the past becomes a problem for contemporary representation. In this case, we would have no shortage of likely suspects, from the popularization of that familiar affect "hope" in the twelfth century, to the socialization of a compensatory futurity in Christianity—in which, in Pope's famous words, "Man never is, but always to be blest" (I.96)—to the teleological idea of progress in general, one that jumps ship from the progression of spirit (the Hegelian *geist*) to science, sometime in the nineteenth century. However, at the same time it seems that the intense focus on the future that marks the present is very much a modern phenomenon, one that seems to take off in the twentieth century, which, as Franco Berardi has recently suggested, seems to have found its "cultural and ideological inauguration" in a certain kind of obsession with the future, from its aesthetics (most obviously in Italian Futurism, but also elsewhere), to the acceleration of collective trust in the ability of science, technology, and communication media to transform the world at an ever more rapid pace (17). Thus, while I will have cause to return to the longer history of thinking about the future below, I want to start here by focusing on three manifestations of "nostalgia for the future" that are more timely, and thus more significant, for any attempt to periodize the present. Taken together, they suggest how a shifting psychic investment from an abstracted past to an abstract future might be one of the best ways to demarcate the eclipse of the postmodern and emergence of some other cultural dominant on the horizon.

First, we might gauge such an investment, quite literally, in economic terms, through the growth of futures markets. If Jameson's nostalgia for the present during "late capitalism" marked the collapse of distinctions between culture and economics, one seen in the increasing centrality of advertising and the general commodification of aesthetic representation, a nostalgia for the future is marked instead by finance capital and "futures markets": the trading of futures, forwards, and derivatives, in which the present value of a commodity is paradoxically indexed by wagers on its presumed future value. The massive growth of futures markets over the past few decades is certainly notable for its size; according to some estimates, the notional value being traded in derivatives reached $1.2 quadrillion in 2010, or twenty times the "actual" world GDP.[1] However, what might be most striking is the strange intensification and expansion of the commodity form and of speculative capital that has made such increases possible. Indeed, futures markets almost seem explicitly designed to overcome the two limits traditionally posited for capitalism's growth. The first of these is the limit of available market territories, the necessarily limited amount of "space" and resources available for capitalist expansion. Marx famously devoted the final chapter ("The Modern Meaning of Colonization") of the first volume of *Capital* to this dilemma, an examination of colonization schemes promoted to relieve the social antagonisms caused by overpopulation in Britain. Such schemes, for Marx, illustrate the pressure put on capitalist management to assuage potentially revolutionary unrest, but also reveal how these strategies end up reproducing the very same antagonisms on a new plane. Thus "the secret discovered in the New World by the political economy of the Old World" are the same contradictions Marx has already identified, and expansion into new territories is at best a deferral of the crises endemic to capitalist expropriation: there is only so much virgin territory available to be despoiled by capitalists, and the advent of the world market also spells the beginning of the end for a capitalist expansion that burns ever faster through a shrinking amount of unexploited lands (940).

The other limit is also capably conceptualized by Marx in an extension of the M-C-M' formula of capitalist exchange referred to throughout *Capital*; accumulated wealth (M) is invested in commodity production to become capital (C), and those commodities are then sold to produce profit (M'). The problem for the capitalist, however, is to find ways to shorten or eliminate the intermediate step in this formula, to contract the M-C-M' process to an M-M' one wherein money need not be transformed into a good or service before returning back the capitalist as (more) money, to directly realize what Marx calls "money which is worth more money, value which is greater than itself" (257). The massive growth of futures markets and derivatives, and, in particular, the selling of increasingly complicated "financial products" composed by bundling portfolios of existing debt and risk, seem to have solved both of these problems in a single transaction—or, we might say more precisely, infinitely deferred them. On the one hand, if globalization and the rise of the multinational corporations seemed to suggest an exhaustion of the spatial limits of capitalist expansion, then futures markets suggests its inevitable extension through time. On the other hand, products created around speculative finance are as close as one might get to making a commodity out of "money" itself, of realizing invested capital's "future value" in the present.

We might locate another manifestation of this trope in the recent "turn to ethics" in contemporary philosophy and critical theory, at least insofar as we might grant that these discourses, though themselves meant to be *diagnoses* of contemporary social reality, can also be read as *symptoms* of the same. Although there are certainly a significant number of important differences between the approach to ethics by such figures as Agamben, Badiou, Derrida, Hardt and Negri, and Zizek, they all share a consistent focus on ethical subjectivity and action as particularly ideal, scarce or undeveloped in the present, but seemingly on the horizon of the (near or distant) future. The idea of a potential collective ethical subject might be the stronger emphasis in Agamben and Hardt and Negri (through their respective concepts of the "coming community" and the "multitude"), while a potential future moment of ethical realization is the stronger emphasis in the works of Derrida (via his writings on such concepts as "justice-to-come") and Zizek (through his focus on the potential ethical acts of the "leap of faith" or "traversing the fantasy"); for his part, a combination of both these figures seem to be the central interest of Badiou (in his writings on what he calls the "subject to the event"). However, for all of these writers, ethics is typically approached by way of how ethical subjectivity or community *might occur* in conditions other than those of the present. Further, insofar as these writers also tend to position their hopes for a more ethically positive future against the vagaries of contemporary capitalism as a social dominant (of which, more in a moment), we might say they form a parallel and opposing "futures market" of ideas about ethical possibility against that of capitalism.

Finally, we might consider recent political contexts, in the American scene at least, and how arguments have tended to insist on a certain optimistic depiction of the future in place of concrete plans for present action. In addition to the well-known emphasis on "hope" in the 2008 Democratic presidential campaign, we might also consider, in this regard, former President George W. Bush's repeated insistence that despite the negative appraisals of his performance that were current at the end of his presidency, his actions will be judged favorably when they are recounted in the "history" of the future. For that matter, we might also take frequent suggestions made on behalf of those opposing new regulations based on ecological concerns that an increasingly compromised environmental ecology is actually in "better shape" each new day, because advances in science and technology are improving our chances of solving ecological dilemmas and the crisis of global warming.

In this chapter, I take up a technological and rhetorical ecology that might at first blush be seen paradigmatic of some of the worst excesses of such a "nostalgia for the future" (particular its economic logic), but that I will suggest might offer productive strategies for responding to the ethical and political challenges of the present moment: the possibility of "material immortality," the belief that radical life extension may be realized in the (near or distant) future. Although there are a variety of ways in which the possibility of ending natural (causes of) death—from the acceleration of existing medical and technological practices for life extension, to the discovery and reversal of the aging process in humans, to the "transference" of human consciousness or subjectivity into some other biological or technological realm—I am ultimately interested in how the concept of material immortality itself might literalize or make immanent the central wellspring of Western ethical thought, our originary ethics. Although based on a very different conception of immortality, the Platonic "creation" (or revision) of the soul and theorization of ethical subjectivity, I will suggest here, still very much occupies the center of contemporary ethics of all stripes even if only as, in line with traditional notions of the soul "itself," an uninvoked but animating spirit behind ethical thought, valuation, and judgment.

I begin, below, with what we what is often taken to be the "primal scene" or originary articulation point for a variety of disciplines and divisions that I have been, in many ways, trailing throughout this text: the birth of Platonic philosophy and Western ethics as it becomes defined against sophistic rhetoric in ancient Greece, one that, as we have seen, involved a more general division between the categories of *logos* and *techne*, and one that, as I will emphasize here, also takes place shortly after the emergence of formalized commodity exchanges and the development of a standardized monetary system (thus marking the birth of the modern forms of both *ethics* and of *economics*). Although there are surely a multitude of interesting historical parallels and divergences that might be covered in these two sites, I am specifically interested in leveraging the metaphysical legacy of "originary" Western ethics and the more current "(re)turn to ethics" in critical theory, and using this swerve through Greek antiquity to foreground the different ways critical discourses of ethics have been, and possibly could be, positioned in relation to changes in both media technologies and contemporary processes of economic commodification. In the end, I'll be arguing for the contemporary potential of a "sophistically" indebted conception of ethics that prioritizes changes in human activity over changes in human subjectivity. All of which is to say, though I will be initially focusing on a centuries-old dispute, I am ultimately concerned with the question of today: our possible ethical and rhetorical responses to the demands of the present moment.

THE SOUL OF THE FIRST CONSUMERS

> If I actually had a soul made of gold, Callicles, don't you think I'd be pleased to find one of those stones on which they test gold?
>
> —Plato, *Gorgias* (486d)

> As the speed with which new and original products are transformed into commodities reduces, those that can most easily be reproduced rapidly fall in value, while those with genuinely authentic features hold or even increase their worth.
>
> — David Lewis and Darren Bridger,
> *The Soul of the New Consumer* (200)

Insofar as this chapter is an attempt to try to think ethics, and in particularly its pragmatic intersections with politics and economics, without or *beyond* the ontological subject as a locus of agency and ethical reasoning developed post-Plato, I want to pursue the different conclusions and strategies that might be gained by revisiting the emergence of both commodity capitalism and the ethical subject in the West. If, as I have argued several times throughout this book, the challenges of the contemporary moment are very much bound up with confronting the inseparability of capitalist desire and human desire "itself" and the related importance of rhetoric or the aleatory forces of *suasion* as the modern inheritor of the role formerly occupied by ideology or identity, then returning to primal scene through which Western rhetoric, ethics, and commodity capitalism emerged would seem to be a particularly privileged location for staging such an intervention.

Indeed, although much current work on ethics may emphasize the contentious relationship between contemporary capitalism and contemporary ethics, as Georg Simmel notes in *The Philosophy of Money*, the broad domains of economics and ethics have always possessed a certain methodological overlap; as disciplinary formations or simply fields of forces in practice, both produce "the objectification of subjective values": the formalization of procedures for determining the value of tangible and intangible phenomena based on precedent and/or context, whether they be the market price of a commodity or the moral "worth" of an individual or behavior (65). Although, again, Plato's work is often read as resolving a particular tension between these two fields—as positioning ethical reason as *either* an alternative to economic reason *or* as the standard of judgment through which economic systems and relationships might themselves be judged—I am interested in what might be lost in drawing such conclusions, particularly as they relate to our contemporary tendency to position the desires unfolded by capitalism as an economic system in opposition to the more "high-minded" forms of collectivity or sociality valorized within ethical theory. In this sense then, what is at stake in the following is not so much a consideration of how this moment provides access to the early history of the entwined evolution of ethics and economics, but rather how it may reveal the mutual emergence and co-dependence of them as we understand them today: the extent to which ethics, as we have inherited it as a concept, quite literally comes into being as, on the one hand, a parasitically mimetic appropriation or inheritance of the formal abstraction demonstrated in monetary exchange, and, on the other, a consistently critical rejection of the modes of subjectivity and sociality that seem to be engendered by that same process. To what extent is our ability to imagine ethical action and the relationship between ethical and economic exchange constrained by the legacies of the moment in which both capitalism and ethics materialize in Western culture?

As alluded to above, of particular interest here will be the Platonic "invention" of the soul and of an ethical conception of subjectivity and the ways in which Plato's theorization of these concepts is worked out against the emerging monetization of Greek culture. As Nietzsche's contemporary Erwin Rohde was one of the first to acknowledge, Plato was largely responsible for conceiving what would become the homodoxical view of the soul in Western thought up until the present and in many ways also our understanding of the "subject" as a self-reflective being; though *psyche* and related terms circulated as amorphous figures of animal "life force" or the ostensible cause for various human capacities, it was Plato who seems to have popularized its definition as a malleable vector of subjectivity that affects, and is affected by, its possessor's mentality, moral judgment, and behavior (463–476). Though far less commented upon, Plato's references to money and the burgeoning commodities markets of Greece are equally pervasive. Most striking given my purposes here is the high frequency in which consideration of the soul and of money occur together with the value of one placed against the other, as in the epigraph at the beginning of this essay or, for instance, the famous passage in *The Republic* where Plato suggests the guardians of the citizenry will abjure material wealth because they have gold and silver "of the divine sort" in their souls (416e).

Indeed, Plato's tendencies to analyze material wealth and the use of money as a negative counterpoint to the worth and use of virtue performs a large degree of heavy lifting in service of two of Plato's most novel philosophical objectives: (1) a revaluation of traditional notions of justice and virtue during a time of quasi-secularization and the extension of democratic governance, and (2) a not-unrelated critique of the effects of increased commodification in Greek life, a force that was equally disturbing to both the aristocracy and theologically or mythically indebted systems of the social order. While the latter objective is much less celebrated in references to Plato's legacy, it may in fact have been the more radical notion during Plato's time. Although it may be difficult to imagine now, as M. I. Finley writes, up until the late fifth century the "judgment of antiquity about wealth was fundamentally unequivocal and uncomplicated. Wealth was necessary and it was good" (35–36; cited in Shaps 131). Among other contributing cultural factors that to led the problematization of wealth, notably the influx of foreign capital from outside of Greece, was the advent of the Greek coinage system. Coined money disrupted traditional social and economic practices based on the accumulation and inheritance of property and prestige commodities, making it possible for the ill born, even slaves, to potentially accrue and store wealth. The sea change brought by the combination of foreign capital and coinage on Greek social hierarchies and systems of valuation (both ethical and material) amounted to, as David Shaps writes, "a moral economy turned on its head" (133).

In this sense, then, it might be more accurate to consider Plato's critique of wealth, and his forwarding of a new ethical system through the concept of the soul, as not so much overlapping endeavors as they are a single gesture through which the "first philosophy" of Platonic metaphysics emerges. More specifically, the comparison between "value" in monetary and ethical contexts as it appears in Plato's philosophy can be profitably compared against the philosophical commonplaces that preceded it. As numerous classicists have documented, the monism of the pre-Socratic philosophers is consistently linked to the outbreak of formal monetization in surviving writings of the time. Thus, as Richard Seaford suggests, pre-Socratic metaphysics was at least partially a "cosmological projection of the universal power and universal exchange of the abstract substance of money," its status as an "impersonal all-powerful substance," that disrupted traditional Greek systems of valuation and social status (11).[2] Against this backdrop, the Socratic/Platonic dualism organized around the soul as a personal, all-powerful substance superior to economic measures of monetary value reads much like an attempt to restore hierarchies and standards of evaluation around an ethics openly suspicious of, or even hostile to, monetary wealth and the desires associated with its pursuit.

This hostility, however, is not really focused on money itself—one can't really be "for" or "against" money in the abstract, and neither is Plato, that pro-aristocrat conservative, particularly disturbed by the inequality of individuals as measured by material wealth. Rather, we might even say that it is "money" (or the outbreak of formalized abstract exchange on a communal level) that makes Plato's (re)invention of ethics at all possible. If money presented the Pre-Socratics with a positive example of a potentially infinitely unbounded system of abstract exchange and adequation—the flow of money anticipating and inspiring a Parmenidean "One Being" or Heraclitian fire—it also became the necessary background through which Plato could construct an alternative system of valuation that would position such an exchange to be an artifice obscuring our understanding of "real" value.

As such, and this is where my provocation around Plato's views on monetization takes on more than a mere historical interest, monetary value becomes a crucial part of a constellation of concepts or forces that gives birth to Western ethics and, in particular, the suturing of ethics to the work or effect of an ethical soul or subject, that individual who has understood and committed themselves to the correct values (and value system). Plato's invention of the "ethical subject" in this sense emerges out of his triangulation of (1) new forms of valuation taking place in the monetary economy; (2) his recreation of *psyche* as both the "moral" soul and cognitive seat of judgment in the individual that would serve as a counterforce to economic or populist valuation; and, (3) particularly importantly for my purposes here, his critique of the sophists, the loosely connected group of itinerant rhetoricians and pedagogues whose few constitutive qualities included the charging of money for their services.

Although I trace the specific relevance of this the transaction for shaping our subsequent understanding of ethics, we might take up the dialogues *Meno* and the *Sophist*—two works that are themselves very much occupied with the question of the representative, the authentic, and the synecdochal—as exemplary of this sequence as it occurs in part or in whole throughout Plato's works. In both dialogues, Plato emphasizes the difference between the dialectical division that would allow one to recognize virtue and to act in an ethical manner from that of monetary exchange or economic calculation, emphasizing the errancy of those who, as Marc Shell writes, "unwittingly divide up the conceptual and political world in which they live by a kind of division that is formally identical with money changing" (131). In both, Plato also contrasts his own consideration of virtue and how it appears through philosophical reason with that of the sophists and its use in rhetorical praxis. Finally, in both cases, Plato introduces the soul as a conceptual abstraction that allows one to "quantify" goodness or human virtue in a way that mirrors the rigor of economic exchanges even as it attains its identity in contrast to that domain.

The import of these distinctions, however, comes through most strongly in Socrates' interactions with his sophistic competitors. More specifically, as presented in the dialogues, Socrates' encounters with sophists reveal fundamental contrasts between sophistic and Socratic/Platonic thinking on three question central to ethics, ones that may seem overwhelmingly "decided" today, given the long shadow Platonic ethics casts on the Western tradition, but which appear to be much more fluid during the time of Plato's writing. Given the Socratic/Platonic depiction of the sophists as mercenary figures as well as the late twentieth-century rehabilitation of the sophists as prototypical thinkers of various types of relativism, it is not surprising that the sophists are not often taken to have any sustained interest in ethics.[3] However, even in the Platonic dialogues, no fewer than three of which depict Socrates debating sophists on questions of virtue and whether virtue can be taught, the sophists are shown actively defending their interest in ethics and its importance in the populace. Lack of attention to the particular ethical thinking of the sophists in these dialogues is, I will suggest below, not so much the result of absence of attention to these issues on the part of the sophists; rather, Platonic definition of ethics has come to so forcefully dominate our conception of what can be counted as ethics, that it is often quite difficult to recognize the sophists' ideas as competing ethical theories. We might trace the disagreement between Socrates and the sophists, as well as the "decisions" that would guide thinking about ethics since that time, around the following three core questions.

Can ethical behavior be both selfless and self-interested?

In the opening of the *Protagoras*, Socrates and the titular sophist of the dialogue embark on a long debate over the nature of pleasure and pain and Socrates admonishes the sophist for practicing a rhetorical pedagogy that is self-indulgent for both parties, meant to be pleasurable for both student and instructor. If their practices have an ethical or altruistic component— and of course the sophists do argue that they are ethical and improve the community they address—Socrates argues that such a selflessness emerges only, paradoxically, as a corollary to their own selfishness, particularly their desire to increase their material wealth, their professional reputations, or their image as wise and respectable individuals. Socrates' indictment of sophistic practice in this regard is twofold. On the one hand, since the sophists charge money for their services, they are indifferent as to who might become their students (in the dialogues, sophists repeatedly advertise their willingness to accept any student that can pay their fee); as Socrates tells Hippocrates, "if you meet his price he'll make you wise too" (310d). Similarly, they are also beholden to satisfying the students' extant desires for their education (thus sophists will enroll students of various moral calibers and such students are likely pursuing education for the mercenary goals of gaining profit and power). On the other hand, since the sophists make great claims for both the ease and effectiveness of their training, they must demand little sacrifice or risk on behalf of their students, qualities Socrates considers necessary for leading an ethical life. Socrates criticizes all of these qualities and their supposed dangers early in the dialogue, warning Hippocrates that while the sophists "take their teachings from town to town and sell them wholesale or retail to anybody who wants them" and "recommend all their products," they do not know "which are beneficial and which are detrimental to the soul" (313d).

After a similar consideration of pleasure and pain in the *Gorgias*, rhetoric is compared to a type of flattery wherein the rhetorician derives pleasure from reinforcing the desires of his audience, and philosophy is compared to the administration of harsh medical treatments that are painful for both the practitioner and the patient (463a). In both cases, Plato introduces the distinction between the body and an immortal soul to distinguish these practices and to condemn rhetoric, as he does also in the *Meno* where the immortal souls is introduced to convince Meno to give up what Socrates calls "the debater's talk" of rhetoricians (80e), and in the *Apology* wherein Socrates catalogues how he has sacrificed mortal pleasures to convince others to take care of their immortal souls.

For their part, Gorgias and his allies, specifically Gorgias' compatriot Callicles, upbraid Socrates' for being hypocritical in his celebration of selflessness and his ostensible disinterest in reputation economies or strategies of self-presentation. Anticipating Nietzsche's later critique of Socrates as "full of ulterior motives" and "the cleverest of all self-deceivers," Callicles argues that Socrates' presumed selflessness and disregard for pleasure is only a mask for a particular type of selfishness and pleasure-seeking (that of being "right" and ethically superior to others) (*Twilight* 4, 12).[4] Inverting Socrates' critique of the sophists' practices and ideational investments, Callicles claims Socrates himself is guilty of indulging in "crowd-pleasing vulgarities" (584e), is cowardly because he spends "his life in hiding" from material concerns (482e), and, despite his protests to the contrary, does indeed "love to win" in arguments and not simply seeking to reveal the truth through dialogue with others (515b). We might contrast these two approaches in reference to how they attempt to contrast or expand the field of "ethically positive" action. For Socrates, even if we can agree that a particular outcome or effect has served a positive ethical value (that of serving justice, or increasing virtue in the populace), we must also interrogate further whether the actions behind that effect were ethically pure, or more precisely, to determine whether the motivation behind whatever action led to that effect was in fact a selfish one (notably the gaining of wealth or esteem). The sophists, however, counter that all motivations presume in the one way or the other a certain gain, even if it is the self-satisfaction of "doing good," and although Socrates might indict them for accepting money for their teaching, he too receives payment of this sort. All of the disagreements between the sophists and Socrates as represented in the dialogues might be read as extensions of this central conflict. This contrast between the selfish or self-serving vectors of ethical judgments and actions (indicted by Socrates while affirmed, and suspected in others, by the sophists), may be the best way to discriminate the ethical dimensions of Platonic philosophy and sophistic rhetoric, contra more typical references to the primacy of "truth" versus "persuasion" or normative versus situated approaches to ethical reason.

How does one evaluate the success of an ethical program or praxis?

Plato's denigration of the immediate benefits and pleasures promised by sophistic training also led to a reconception of the timing, as well as the measures, of evaluating ethical actions. It is, of course, not altogether surprising that Plato's Socrates would generally refrain from trumpeting the success of his ethical system; this is, after all, the guy who claimed anyone "who really fights for justice must lead a private, not a public, life if he is to survive for even a short time" (*Apology* 32a). However, against the sophist's entrepreneurial pitch for the "immediate results" of their actions on both students and the populace at large, Socrates advocates the success of his ethical training on its potential future, or indeed, its potentiality *tout court*. For Socrates the production of ethical action—acting virtuously on the level of the individual or the social—by necessity can only be achieved by the ethical subject: the individual who has learned to take care of themselves and what virtue "is" (or at least what virtue is appropriate for the doer). Although action that might appear virtuous (and Socrates has no end of examples of same) might occur frequently, they are not truly so if the subject performing them has not understood virtue. This is why the presumed fomenting of virtue by the sophists is consistently indicted—they, and their followers, know not what they do.

Since virtue in the objective world can only occur through individuals who have made such a subjective change and, as Plato never tires of reminding us, such individuals are in short supply, it is not surprising that the success of any Platonic revolution in values would not be immediately forthcoming. The potential for "real" ethical action, however, is always on the horizon, the promised result of the sea change that would take place if the number of individuals who learned to "take care of" their souls reached the point where they could transform the *demos*, or a select few, such as the titular proto-politician of the first *Alcibiades* or the prototypical "philosopher-kings" of the *Republic*, would assume political power. The most vivid illustration of the Platonic deferral of evaluation emerges, not surprisingly, in his depictions of the "final judgment" of the immortal soul after death. In the famous concluding monologue to the *Gorgias*, Socrates distinguishes how such judgment took place in the previous "time of Kronos" versus the contemporary "time of Zeus" (a mythology that would be repeated in the "sequel" to the *Sophist*, the *Statesman*). In Socrates' retelling, under the law of Kronos, humans have knowledge of their earthly death, appear for divine judgment in bodily form and with their raiment and wealth, and are able to draw upon the testimony of persuasive witnesses as to their moral wealth. Under the laws of Zeus, however, death arrives unexpectedly, and only the naked souls of individuals are judged and the truth of their being is revealed. Although Callicles' prescient suggestion that Socrates would be unable to defend himself in an earthly court is perhaps the indictment best remembered from the dialogue, Socrates concludes the monologue by criticizing his sophistic interlocutors for acting as if they are still living under the law of Kronos and thus mistaking the importance of their actions and their appearance to others as compared to the "real" state of their soul. As a synecdoche for the Platonic judgment of ethical worth, the parable foregrounds both a deferral of the evaluation of the immediate material consequences of an individual's actions as well as the introduction of an entirely new system of ethical valuation—one in which the motivation or "soul" of the individual is the target of an particular ethical program before and above any of the effects that they might produce in the material world.

Through what means is an ethical subjectivity or ethical program passed on to others?

Finally, in distinguishing his and Socrates' teachings from those of the sophists, Plato needs to account for how ethical values or behaviors can be developed in others. In this endeavor, particularly when directly called upon to provide such an accounting, the Platonic soul would consistently serve as a *deux ex machina* in the dialogues (in a sense, quite literally, given Plato's association of the soul with what amounts to a divine share within the corporeal body). Crucial to this endeavor would be Plato's reimagining of the soul as a vulnerable and perpetually imperiled vector of subjectivity, contra the sophists' near-celebration of the soul's openness to being moved and persuaded. This is not to say that Socrates denies his own influence on the souls of his students; indeed, this is Socrates' "job," his self-professed sole vocation: "For I go around doing nothing but persuading both young and old among you not to care for your body or your wealth in preference to or as strongly as for the best possible state of your soul" (*Apology* 30b). Nor does Plato deny the ability of the sophists to impact the soul of others either; the *Protagoras*, the *Gorgias*, and the *Meno* all show Socrates attesting to the sophists' ability to alter souls. However, Socrates and the sophists have very different conceptions of how to appraise the soul's ability to be affected by outside forces. On the one hand, reference by the sophists to the soul and its ability to affected by persuasion, in their surviving writings as well as in the dialogues, present, on the whole, this ability as fundamentally a positive one; indeed, a cryptic statement attributed to Gorgias by Plutarch provides perhaps the most intense version of this conception: "the deceiver is more justly esteemed than the nondeceiver and the deceived is wiser than the undeceived" (Kennedy 65). The Platonic reframing of the soul and its relationship to rhetoric would flip both the general valuation of the soul's "openness" to rhetoric and the best way for it to be managed. On the one hand, persuasion, particularly the vintage that draws on popular desires and intellectual commonplaces, becomes a constant threat to the soul. Plato's instruction to commit to the beliefs and investments appropriate to one's soul against the vulgar persuasions of the populace at large form the conceptual foundation *and* the *praxis* for his ethics (in the sense of self-formation as well as in regards to the determination of just or equitable behavior).

This, then, is at least the form taken by ethics in Plato's writings. As Iakovos Vasiliou argues in his recent *Aiming at Virtue in Plato*, though Plato consistently affirmed the "supremacy" of virtue in his ethical and philosophical work, he was also maddeningly obscure in spelling out precisely what might be defined as virtue or as virtuous; at best such a question in answered only be deferral to the Platonic dualism of the "Ideal Forms" and material reality (or, as suggested above, the not unrelated dualism of the body and soul) (283). In other words, in place of a discrete definition or series of commonplaces, one is given protocols for ways of "being ethical." As mentioned earlier, such protocols are perhaps shown in the best relief when read through the encounters with the first sophists (who's ethical strategies are very much the inverse of Plato's) and against the backdrop of the emergence of commodification systems in Greek economics and culture. These are also the protocols to which much contemporary work on ethics in critical theory shows a surprising fidelity.[5]

ETHICS ETERNAL, OR POST-POSTMODERN ETHICS

Of course, insofar as the Platonic "invention" of ethics is precisely that—the originary determination of the what "counts" as ethical reasoning or concern—it is not altogether shocking to see its fundamental features continue into the present. Indeed, as I will suggest below, distinctions between different approaches to ethics in the present are perhaps best contrasted not so much by how they diverge or adhere to Plato's work, but rather by how they emphasize particular aspects of the Platonic heritage. What is perhaps most striking and most relevant to the argument under review in this essay, however, are the ways in which the imbrication of political and ethical thinking encapsulated in the so-called "ethical turn" in contemporary theory have largely absorbed or intensified the Platonic side of the three conflicts between sophistic and platonic, or early rhetorical and philosophical, engagements with ethics outlined above. In other words, if the shift that largely put the "post" in "postmodernism" was a rejection or severe suspicion of not only touchstones of Enlightenment thought, but of the Western metaphysical heritage altogether, it would seem particularly strange that Plato's influence, and his particular construction of ethical subjectivity, would return so strongly in thinkers working within and in response to postmodern theory and poststructuralist philosophy.

Indeed, precisely because it seems so out-of-step with these movements, we might say that the ethico-political emphasis of contemporary theorists has formed the first identifiably "post-postmodern" current in critical theory and philosophy; this is particularly the case if we think of ethics in broader terms than those usually encompassed in references to "the ethical turn" in the humanities and social sciences. The latter endeavor is often narrowly defined around the renewed interest in the work of Emmanuel Levinas, particularly in Derrida's late writings, and two broader critical trends partially inspired by this interest: the increased concentration on the ethics of literary and cultural interpretation, and the "post-secular" emphasis on the intersections of theology and philosophy. My wager here is that including recent work on ethics by Hardt and Negri (writing together and separately), Badiou, and Zizek gives us a fuller picture of ethics within current critical theory, its underlying assumptions as well as its genealogical position (what it is a "turn from"), and the shared methods or objectives of the field (what it is a "turn toward"), particularly in regards to how it diverges from traditional hallmarks of postmodern thought. Below I unpack the shared priorities and strategies of work in this area and its salutary effect in reinvigorating ethics as a viable entry point into interrogating the political; however, I am ultimately interested in emphasizing the limitations of this movement for intervening into contemporary social conditions, limitations with strong connections to the Platonic heritage of ethics as we saw it emerge from pre-Socratic philosophy and in opposition to sophistic thinking.

We might begin with the turn to ontology itself, the characteristic that is perhaps the most general point of connection between current work on ethics within contemporary theory as well as the quality that would seem to be most at odds with the early thrust of that endeavor.[6] If the key concept of early work in postmodern theory and poststructuralist philosophy was subjectivity, and in particular its ostensibly "fragmented" or "constructed" nature (a feeling taken to be newly present and urgently felt during the political upheavals of mid-century), then the turn back to the more traditional category of ontology as an investigation into constitutive "being" would seem to be at least in part an attempt to revise some of the more moribund elements of that initial focus (whether one would undertake it in response to some perceived shortcoming internal to the approach, or in response to the changed social conditions taking place around the end of the twentieth century). Although emphasizing the sociality and mutability of subjectivity *tout court* increasingly put pressure on conservative thinkers' recourse to a transhistorical or "essential" view of human beings that could cut across a multitude of material and experiential differences, the unrelenting critique of sociality as "a problem" for the subject might have paradoxically lead to its own presumption of a "natural" and ethical human subject waiting to be freed from societal constraints. In other words, if subjectivity is taken to be constructed socially and the critique of that process becomes the focus of progressive efforts to redress the negative political and ethical effects of identity, we might end up switching sides with our opponents and implicitly presuming, as Judith Butler writes, the existence of "a natural eros that has been subsequently denied by a restrictive culture" and was not itself open to interrogation (*Subjects* 214). Additionally, if, as has been suggested throughout this book, contemporary social power no longer seems to work through the imposition of particular norms and values as much as it does from the extraction of value from a variety of identities and behaviors, then focusing on the instability or flexiblity *of* power would hardly seem to do the same amount of heavy lifting that this kind of critique was taken to perform in an early moment. Thus, in this sense, it might not be altogether surprising that ethics became a crucial category for participants in what Carsten Strathausen calls the development of a "neo-left ontology" geared toward imagining different ways of being (a subject) and of being together (part of a social whole).

For writers such Badiou, Zizek, Hardt, Negri, and Derrida, then, ethics, in both of its ontological and practical forms, has become a crucial site for shifting from a "postmodern" focus on the diagnosis of power and culture to an emergent effort to devise more explicitly prescriptive propositions for political and ethical action. To phrase this distinction somewhat differently, if the "big theory" era of postmodern critical work was beholden to interpretation as an (ethically or politically) performative act—an attack on rigid constructions of meaning or belief that structure social power is taken to be itself an ameliorative gesture, or meant to cash out, however abstractly, as a resource for discrete political action—then "post-postmodern" work on ethics instead configures interpretation as still prior in sequence but secondary in importance: diagnosis becomes a means to the end of crafting prescriptive measures for one's individual and collective activities. More precisely, this shift has largely been one from the interrogation of subjectivity as a concept (its fragmentation, ideological debts, singularity, etc.) to thematizing the appropriate or ameliorative subjectivity necessary to effectively respond to contemporary circumstances (the ethical subject, the revolutionary subject, the subject of truth, the member of the multitude, etc.).

And it is here—moving outward from the (re)turn to ethics and ontology in contemporary political thought—that we can begin to see the odd ways in which even the most "radical" attempts to rethink ethics for the post-postmodern present remain beholden to the premodern commonplaces of Plato's work. First, all of the theorists named above continue the Platonic gesture of forwarding an (ideal) formalism by thematizing ethics as an exemplary operation; rather than proposing particular sets of practices or generic standards of valuation for ethical action, ethics is articulated in their works largely through reference to (often heroic) figures and events that might encapsulate the potential and power of ethical praxis. For instance, despite their many other differences, Derrida's and Badiou's shared conceptions of the singularity of the ethical subject and of ethical "events" leads them both to focus on exemplary figures that might stand in to evoke the "(im)possibility" of ethical decision-making and responsibility (Abraham for Derrida) or represent a subjectivity driven not by its constitutive identity but its ethical fidelity to an event or "truth" (St. Paul for Badiou). Similarly, for Hardt and Negri, since the revolutionary community of the multitude is by definition "composed of innumerable internal differences," they draw on specific individuals (St. Francis), groups (the Zapatistas), and events (the WTO protests in Seattle, the Tianemen Square protests, the 1992 L.A. riots) that suggest possible forms of community that preserve such internal differences while still promoting a collective ethical and political identity

(*Multitude* xiv). The use of exemplary figures has also, of course, become a signature move in Zizek's work, whose writings consistently forward a variety of fictional and real individuals—from Sophocles' Antigone to Toni Morrison's Sethe to student-seducing former American schoolteacher Mary Kay Letourneau—who have achieved Lacan's ethical maxim to "not give up on your desire" or have learned to "traverse the fantasy."[7] For Zizek, as for the others mentioned above, ethics must be tied to particular practices to have any instructive import, but since the ethical evaluations of these actions themselves are dependent on an understanding of their "subjective" circumstances, and indeed the very subjectivity of the individuals performing them, ethical action is best assayed through such exemplary figures and sites. To use Badiou's vocabulary, exemplary forms are a salutary way to demonstrate ethical action as both "universal"—open to being understood and adopted by others—and "singular"—taking place in reference to particular objective and subjective circumstances.

Second, repeating in many ways Plato's positioning of ethics against the emergence of the formal commodities market, all of these theorists index the potential and performance of ethics against processes of commodification under contemporary capitalism. Although it may be a given that any contemporary theorizing of ethical or political action must, for good or ill, be thought against the backdrop of capitalism, for all of the above, ethical action is largely evaluated in reference to its capacity for operating "outside" or counter to commodification as a self-interested of profit-seeking operation. We can see this vector, for instance, in Derrida's consistent interest in the concept of "the gift" and in pre- (or proto-)capitalistic gift economies; although present in his work from *Glas* onward, the gift becomes increasingly important in the later works on ethics such as *Given Time* and *The Gift of Death* as an action that "supposes a break with reciprocity, exchange, economy, and circular movement" and might be leveraged for thematizing larger notions of ethics, hospitality and responsibility ("Hospitality" 69).[8] For Badiou, the organization of social life around "money as [the] general form of equivalence" and "commercial value" as the only recognized value, undergirds what he calls the "nihilistic ethics"—comprising both philosophical discourses of "otherness" and populist notions of human rights and humanitarian intervention—to which his proposed "ethic of truths" will be a counterpoint (31). Zizek takes a similar angle in identifying "the unbridled commodification of everyday life" as the key triumph of contemporary capitalism and the background against which his work on ethics and religion is deployed (*Tarrying* 216). Changes in the logic of capitalist commodification, and in particular of "that special form of commodity that is labor-power" also provide both the challenge and possibility of Hardt and Negri's multitude (*Empire* 397); on the one hand, the increasing importance of immaterial labor, the extraction of value from practices of creativity, knowledge, and communication that would not traditionally be categorized as labor or circumscribed by the work-day, hails the introduction of wholly novel forms of labor exploitation and commodity circulation: "at the pinnacle of contemporary production, information and communication are the very commodities produced" (298). On the other hand, the location of labor value in the creative and affective capacities of the worker also signals a greater control by the worker over the "means of production," and finding ways to reappropriate this value and its circulation becomes "a primary determination of the multitude" and "the first ethical act of a counterimperial ontology" (363).

The positioning of capitalist exchange here is, as it was in Plato, in many ways a synecdoche for the ways that work on ethics within left-oriented theorizing positions ethics as a scarce phenomenon within contemporary life or, we might even say, as wholly unrealizable under present conditions.[9] In the work of Hardt and Negri and of Derrida, we can find this tendency through the oft-noted messianic overtones of their work, the ways in which their shared emphasis on the "virtual" or "immanent" nature of particular ontological forms and ethical possibilities is coupled with a necessary deferral or postponement of their arrival. In Derrida's work, this is perhaps clearest in the focus on the temporally paradoxical designation of the "to come," a phrase he appends to a variety of terms and concepts (democracy, justice, philosophy) to mark both the radical alterity or unknowability of the future as well as the necessity of pursuing ideal notions of such concepts while knowing that they can never be perfectly realized.[10] For their part, Hardt and Negri follow a similar formulation in describing the ontological community of the multitude. On the one hand, the multitude, as a new form of collective sociality that would provide the ontological basis for progressive political and ethical action, is "latent and implicit in our social being" (*Multitude* 221). At the same time, however, it remains unrealized in the present, marking only the pure potential or possibility for its future realization. While they argue for the existence of what they call an "always-already" multitude immanent in contemporary sociality, they are at the same time waiting on what they term the "not-yet multitude": the deferred realization of its powers.

Deferral also remains an important part of Badiou's and Zizek's approach to ethics, though we might better describe them as more specifically marking the rarity of ethical subjectivity by more directly and narrowly associating it with world-historical change or revolutionary events. Despite significant divergences on a number of other points, the two share an approach to bridging the ontological and the political that is perhaps best summarized by the title of Simon Critchley's Levinas- and Badiou-influenced work *Infinitely Demanding*. On one side, the subject is (to use Badiou's terminology) "riven," "called," or even "constituted" by a particular event or course of action; on the other, their concrete involvement in the political then, or at least the involvement we are aiming for, emerges in a necessary opposition to the "natural" or apparent conditions of possibility in the contemporary social environment, a demand that exceeds what is offered by the dominant institutions of social power.

In this sense, as Adrian Johnston suggests in his excellent analysis of Badiou's and Zizek's theories of political transformation, it is useful to consider their work on this question in relation to the old saying often associated with the protests of May '68, "Be Reasonable: Demand the Impossible!" Although the increasing co-option of postmodern or post-structuralist strategies of critique and resistance by dominant institutions of social power would seem to make radical and revolutionary change ever less of a possibility, "Badiou and Zizek tirelessly remind their audiences that conceptions of realistic possibilities are themselves historically transitory constructions" (xvii). Such a focus subtends their interests in unpacking various logics of identity and rupture, as well as their shared emphasis on, if not outright fetishization of, the power of revolutionary moments whose historical rarity is inversely proportional to their refiguring of ethical potentialities.

Although this blend of the metaphysical and the political has for many had the productive effect of explicitly (re)emphasizing how the work of critical theory or philosophy does or should intersect "actual" politics, as many critics have pointed out, it also seems to ignore the incremental, quotidian, or pragmatic vectors of political action in favor of the revolutionary, the quirky, and the ideal. In other words, by presenting the relatively rare revolutionary event and the equally exceptional subjects committed to these events as its privileged examples of political change, Badiou and Zizek often seem to neglect the process through which people are motivated to take part in such actions, as well as the steps that must occur between the static present and the hoped-for future. In this sense, when radical transformation is not actively happening, Badiou's and Zizek's ontological engagement with politics can also look much like an advocation of spontaneous commitment (voluntarism), non-engagement (quietism), and the relentless critique of strategies and movements that fail to meet the rigid criteria of what "counts" as change (absolutism).

There is of, course, much to recommend all of these contributions to the contemporary rethinking of ethics as a whole, not the least of which is their achievement in refining and retuning critical theory around more explicitly political goals. At the same time, however, it is also worth asking after whether the contemporary turn to ethics via ontology has not so much avoided as repeated and intensified what we might call the "elitism" of ethics in Plato's writings. On the one hand, the "post-secular" designation attributed to many of the thinkers under review here due to their engagement with traditional religious texts might be more appropriately thought of as evincing "pre-religious" qualities; the "ethical subject" at work in their writing has much more in common with the "ethical soul" or *psyche* of Plato than it does with subsequent, more determinatively "religious" treatments of the concept. Still further, the emphasis on the deferral of judgment and change, the conceptual "afterlife" of ethical thinking in this register, seems to bear much more in common with the Plato's invocation of immortality as the realm "above" immediate material circumstances and the common desires of the populace. Finally, and perhaps most pragmatically, in a time like the present wherein its is quite difficult to imagine what something like direct "resistance" or "opposition" to contemporary structures of social power might look like, it seems questionable to say the least that conceiving ethics over and against the common desires and ubiquitous material exchanges of the contemporary social sphere would be our most promising, or most efficient, way to mobilize ethics today.

In a continuation of this book's interest in thinking together contemporary forces of *technics* (the impact and operations of contemporary technoscience and technology) and those of rhetoric or social *suasion*, for the remainder of this chapter, and thus the winding down of this book, I want to, first, posit whether a different conception of (im)mortality than that of either Plato or (post-)theological thought might be a better conceptual foundation for working through questions of contemporary opportunities for ethical action, and then, second, attempt to do just that using strategies that owe much to the sophistic thinking about ethics that was largely dismissed in the wake of Plato's ascension. Both of these endeavors will also provide a final opportunity to put a finer point on the sociotechnical changes and ethical and rhetorical practices that I have been organized around the "transhuman" here in this text.

LIFE ON THE INSTALLMENT PLAN: BIOPOLITICS, CAPITAL, AND THE ETHICS OF (IM)MORTALITY

> Man has lost his *soul*; in return, however, he gains his *body*.
> —Georg Lukacs, "Thoughts Toward an Aesthetic of Cinema" (16)

Thus far I have suggested that the surprising similarities between the "premodern" ethics of Plato and the "post-postmodern" ethics implicit in the contemporary turn to ontology emphasize the boundaries to our abilities to imagine what counts as "ethics," particularly our dependence on an ontological conception of an "ethical subject" as the locus and arbiter of ethical activity. If the fidelity between recent work on ethics and the Platonic starting point of Western ethics as a whole suggests a form for ethics that appears transhistorical, the presumed failures of postmodern social theory to account for new forms of social power has, at the same time, had a considerable impact in emphasizing the ontological condition of human beings and human collectivities. One way, then, to interrogate the specific question of the relationship between (the metaphysical category of) ontology and (the practical challenges of) ethics and to begin thinking of an alternative to contemporary ethics that emphasize ontology might be to turn to the historicity or "periodicity" of ontology itself.

As detailed above, despite our tendencies to read Plato's work on ethics as something akin to a branch of his thinking on transcendental forms in general, the use of examples, comparisons, and positioning of the "system" of virtue internal to his writings evince the impact of a number of cultural forces prominent in his time: the "demythification" of Greek thought, the rise of the Greek commodity system, the threat to aristocratic privilege created by both commodity capitalism and democracy, and the larger changes in systems of material and moral valuation marked by Plato as being proper to the "reign of Zeus" versus that of Kronos. These forces not only structure the specific contours of Plato's intervention into moral reason but are also in many ways what make it possible. On the one hand, Plato's writings on virtue are very much positioned as oppositional to the forces he takes to be the most dominant within his own society, those most apparent and influential to the average individual of his time. On the other, however, we might also say that it these very forces that enable him to create alternatives that mimic or hijack the power of the force being opposed, the most important case being Plato's fashioning of dialectical exchange against capitalist exchange, and the forwarding of the body/soul and common/ideal dualisms as both a corrective to the monism of the Pre-Socratics *while at the same time* also being Plato's appropriation of the logic of commodification into philosophical reason. The mediator between these various realms can be found in Plato's theorizing on immortality as the zone occupied by the soul; this realm becomes both a spatial metaphor for the "other world" of the ideal that is eternal or truer than the blandishments of contemporary reality, as well as a zone of futurity itself, a space of deferral within which ethical judgment and evaluation will take place and the "time" during which the spiritual and material effects of Plato's ethics will come to fruition.

Post-postmodern ethics certainly seems like a return to many of these same conditions, or at least of the same dynamic in relation to futurity and ethics. At first blush, we might see the return of something like Plato's ethics in conjunction with a critical rethinking if not outright rejection of an intellectual movement (postmodernism) that was itself defined by the rejection of metaphysical reason that often went by the proper name "Platonic" (not to mention the concept of the "origin" altogether). However, we might more appropriately designate it as the response to a much more modern set of conditions or at least a return to a much more recent engagement with contemporary ontology (and thus, at the same time insist on the historical rather than transhistorical nature of thinking about ontology). More specifically, the turn to ontology in contemporary ethics seems to return us to a series of concerns and priorities that dominated so-called "existential" philosophy of the mid-twentieth century that immediately preceded the emergence of postmodernism as a recognizable movement in philosophy and social theory. In particular, the emphasis in this work on the mortality of the human body and the vulnerability of the human subject seems to play a much more pronounced role than Plato's oft-cited dismissal of corporeality and concentration on the immortal soul. Indeed, these concerns play large not only in the post-postmodern writing on ethics mentioned above but also in the overlapping categories of what Honig calls contemporary "tragic humanism" that we encountered in Chapter 1 and what Carsten Strathausen calls the "neo-left ontology" that forms a common project for contemporary left-leaning political theorists. Here too, however, we might identify a number of historical factors shaping the conception of an ostensibly transhistorical "essence" of human ontology in existential thinking at this time, ones that at least formally parallel those most present in Plato's work: the institutionalization of secularism and waning of religious faith and authority; the growing awareness of the role of language in shaping human consciousness and cognition; and the rapid acceleration of, and intrusion into everyday life, of science and technology as both systems of "reason" and material forces.

However, if the current turn to ontology is guided by very different historical circumstances than its Platonic predecessor and offers more modest claims about the transhistorical or "eternal" or unchanging nature of (human) ontology, the very turn to ontology itself in many resumes Plato's primary emphases on the vectors of ethical thinking we might most closely associate with his conception of "eternity," notably the privileging of the "ethical subject" or individual and the deferral of ethical judgment into the future. As Strathausen argues, while contemporary theorists of political ethics seem to have given up on the possibility of accessing some "Archimedian viewpoint . . . outside the (social or 'natural') space it seeks to analyze," ontology as an analytical category "begins to function as a heuristic device for the historically contingent construction of a different 'nature' from the one we presently inhabit," a way of imagining an ontological condition for future society that might be more amenable to progressive political aims (23). Still further, the emphasis on particular kinds of ontology or the attainment of particular forms of human subjectivity and collectivity proper to these aims brings to mind both Plato's focus on being "appropriate" to one's soul as well as well as what we use to call, in reference, to existentialist thought, the "authentic" subject who has broken free of the common mode of existence enveloping most others and has thus been prepared to act ethically.

It is not at all surprising that in a time of both immense changes in our ability to affect human biology and the parallel retheorization of politics along the lines of the biopolitical described earlier in this book, that contemporary work in ethics would turn to both the material body and as well as the guiding questions of existentialism, the moment of philosophical thought that most directly broached questions of mortality and finitude against the backdrop of earlier accelerations in scientific and technological judgment. What is perhaps harder to figure out, however, is why this movement will end up (re)privileging ontology and the figure of the ethical soul or authentic subject as a solution, a path that would seem intuitively at odds with biopolitical materiality and the function of social power in the "post-ideological" present. As I have alluded to throughout this chapter, I think the answer to this question has to do with the particular boundaries of "ethics" itself as a category, and in particular our inability to think of *ethical activity* in excess of or outside of the agency of the *ethical subject*. Before turning to this question directly, I want to first consider if we might locate a different matrix of sociality and technics in the present that might provide a more productive background for thinking the challenges of ethics today—one that might fulfill the same conceptual function as immortality (of the soul) did for Platonic ethics and the (im)mortality of the human subject performs for more contemporary work in the field, but one that might be both a more pragmatic index of the demands of the present and a more productive backdrop for thinking ethics "after" or beyond the subject.

Specifically, I am interested in the strategic and rhetorical resources we might find in recent considerations of the possibility of "biological" or "material immortality": the notion that human life might be radically extended to a degree in which "mortality" as we traditionally conceive it would no longer be operable for at least some number of human beings. Much like philosophical and theological considerations of the immortal soul or substance or of biological mortality in twentieth-century existentialist and ethical thought, the current consideration of radical life extension seems to lend itself as a synecdoche for the ethical and political challenges of contemporary "human being," particularly those prompted by the impact of science and technology on political economy and forms of social collectivity. More specifically, the key ethical questions posed by the future possibility of radical life extension seem to be more intense versions of those that circulate present-day life. This particular vector of thought about radical life extension is particularly acute not only in the ways that ethical questions are posed by scientists interested in the endeavor, but equally, if not more so, in the wide range of criticisms leveled at interest in the pursuit of biological immortality; in addition to being held in dubious regard by many homodox gerontologists and other life scientists, the very idea of pursuing radical life extension has been roundly critiqued by cultural critics both right and left as one of the most egregious examples of human hubris and our most selfish or self-centered desires (on which, more in a moment). However, it is precisely because of, rather than in spite of, such critiques that I think biological immortality, even if only as a thought experiment, is particularly appropriate site to think through the challenges of contemporary ethics in the same way that other real and imagined states of mortality and immortality have subtended previous iterations of ethics.

Indeed, the variety of different ways scientists have attempted to imagine the possibility or material immortality might naturally suggests such a connection. On the one hand, the multitude of proposals for how "eternal life" might occur currently threaten to equal the number of predictions previously made about possibility of human extinction; on the other, many of these scenarios also seem very much like secularized versions of concepts of the afterlife in theological thought. An obvious example of the latter can be found, for instance, in the work of Frank J. Tipler, a Tulane physicist and whose recent work has involved speculations on the "physics of the future" and how long-term modification of current physical laws might alter the fundamental dimensions of cosmology. In one of these texts Tipler proposes something like a strange repetition or inversion of Pascal's wager; rather than suggesting we act as if God exists based on the smallest possibility that this case may indeed be true, Tipler urges us to act as if God exists based on the idea that humans will eventually create a system of artificial intelligence that will possess all of the power normally attributed to a deity, including resurrection of the dead.[11] Similarly, the prolific inventor Ray Kurzweil has helped further popularize the concept of the "technological singularity" often associated with mathematician and science fiction author Vernor Vinge. In a narrative that has many parallels with millennial or eschatological narratives in Christian thought, Kurzweil has predicted a passage point in the not-too-distant future during which distinctions between humans and machines will break down dramatically and humans will be able to attain material immortality (*Singularity*).

However, considerations of material immortality that are more representative of the endeavor as a whole, as well as much more closely aligned with the contemporary relevance of this question to current questions of ethics, are those that view it as the more "natural" continuation of already existing methods for prolonging and maintaining human life. Proponents of this viewpoint share with the above a certain faith that advances in contemporary science and technology will continue to occur at a pace similar to that of recent decades, but base this belief less on the assumption of some kind of grand shift or event that might make radical life extension possible, but on the more modest assumption that the average healthy life span of humans will continue to expand in parallel with medical advancements and anti-aging therapies; if one accepts this premise, then it is not altogether incredible to imagine that the speed via which the average life span grows will eventually move faster than natural "time" itself—that, at some point, this number will sustain increases of more than one year every year, creating what some proponents have called "actuarial escape velocity," a situation in which an individuals' lifespan could only be predicted as "indefinite."

However, despite the invocation of a certain freedom that might seem inherent in the phrase, in this situation one escapes death only to be trapped by life—life is not so much eternally guaranteed as lived on an installment plan. Thus the key term in "actuarial escape velocity" is not so much "escape," or even "actuarial" (which would suggest a way of indicating individual risk), but rather "velocity," the speed of perpetual motion. From the standpoint of the potential "material immortal," the future looks much like a radical extension of the present, one in which the same problems of contemporary life—notably increased competition for scarce natural resources, global political instability, and the dizzying sociological changes prompted by the accelerated integration of the products of science and technology into everyday life—threaten to sustain and themselves intensify (this is perhaps why from the earliest days of the endeavor, proponents of biological immortality have been drawn to addressing such ethical and political questions as part of their work).[12]

It is precisely this view of the future—as something like a long continuation of the present—as presented in writings about biological immortality that I take to be at the heart of criticisms of the idea rather than, or at least in addition to, more obvious concerns about the sustainability of increased numbers of humans living longer periods of life on earth. Much as other more general questions of genetic manipulation addressed in the first chapter of this book, advocacy for radical life extension have received pointed rebuke from individuals whose political leanings would suggest they otherwise have little else in common. For instance, Leon Kass, the head of the second President Bush's Presidential Commission on Bioethics has vigorously opposed what he calls "the siren song of the conquest on aging and death" as the most dangerous moral challenge posed by contemporary biotechnology, one that encroaches upon traditional religious values and the natural order of biology in deference to the "biomedical gods of good health" and "longer life" (320, 308). Striking a similarly apocalyptic tone, Francis Fukuyama identified "transhumanists" who want "to liberate the human race from its biological constraints" as his pick as one of eight policy advisors asked by the editors of *Foreign Policy* to identify what contemporary idea "poses the greatest threat to the welfare of humanity" ("Transhumanism").

Although, again, one might expect an allergy to this kind of thinking in contemporary neoconservative thought, critiques of the desire for material immortality have also been issued from a very different political standpoint, and one without at least any obvious recourse to the "natural order" of things: contemporary "posthumanist" critical and cultural theory. Carey Wolfe has identified interest in life extension and bodily modification quite succinctly as "'bad' posthumanism" and defines his own sense of positive posthumanism quite directly as "the *opposite*" of the thinking behind such endeavors (*Posthumanism* xvii, xv). To give just one more example, Hayles describes how robotocist and artificial intelligence researcher Hans Moravec's claim that human consciousness could be preserved indefinitely by being "uploaded" into mechanical realms was the "nightmare" that prompted her to write her influential *How We Became Posthuman*, and to argue for a alternative posthumanist perspective "that recognizes and celebrates finitude as a condition of human being" against the possible transcendent subject of an alternative post- or transhuman future (1, 5). In particular, as she has subsequently argued, popular discourses around this question remain "ideologically fraught" with the same "individualism and neoliberal philosophy" that marked humanism and, despite what she acknowledges to be a wide variety of schools of thought around the discourse, all are beholden to visions of the future that preserves "some of the most questionable aspects of capitalist ideology" ("Wrestling"). What concerns Hayles and Wolfe about contemporary populist transhumanism, and what presumably requires them to discriminate their own posthuman(ist) projects from the same despite any number of ostensible similarities, is that transhumanism looks to them less like a critique of humanist notions of autonomy or the tradition of anthropocentrism, and more like an amped up version of the same—a kind of humanism on steroids.

As I have stated throughout this book, arguments for what we might call the "slippery slope" of humanism ("Today: humanism; tomorrow: totalitarianism!") have always been a little too quick for me. And, of course, allegations of the implicit "idealism" of humanism—that it presumes too much about human agency or our ability to rationally figure out the best course for the future—can be just as easily turned around on posthumanism; on many levels, the posthumanist hope that we can "enlighten" ourselves about the role of affect and physiology in human thinking, or the cognitive complexity of animals and the dangers of anthropocentric reason, and through such efforts find a more sustainable way of managing relationships between humans, non-human animals begins to look not all that different from the traditionally "humanist" hope that by putting aside mythological and theological worldviews we might rationally construct a more ethical and humane way of living. And finally, on perhaps the most pedestrian level, it is worth asking after why, with all of the "dangerous" ideas and philosophies currently circulating in global culture, the idea of radical life extension or biological modifications would be singled out as a movement in need of suppression or condemnation?

The answer, I think, lies in two key disjunctions between the view of ethics heralded by considerations of biological immortality as opposed to those presented in homodox Western ethical thinking since Plato's time. On the one hand, while numerous ethical and political theorists have turned to the biopolitical as a crucial realm for analyzing contemporary political economy, this turn has occurred almost entirely by way of emphasizing the manifold avenues through which the normalizing forces of institutionalized politics and/or capitalist production have extended their reach into previously excluded realms.[13] The ethical component of such approaches, when present, tends to come via arguing for the necessary creation of a new ontological condition of human being and collectivity that might resist such encroachment. The question of biological immortality, however, while perhaps only a particularly extreme and only hypothetical example of the same processes often studied under the name of biopolitics, seems to instead emphasize the inherently ethical question of whether life "itself" should be preserved. Gerontologist and life extension proponent Aubrey de Grey, for instance, continually draws our attention to the disconnection between the overwhelming positive consensus around the need to save endangered lives and prevent human-made tragedies as opposed to the relatively common aversion that meets his suggestions for fighting the natural causes of morbidity, the approximately 100,000 people that die daily from "old age" (de Grey and Rae 8–9). Where these two different perspective on the biopolitical diverge most painfully is perhaps around the question of economic interests within life extension—the idea that, since radical life extension would itself presumably be the result of expensive therapies (as life extension for currently ill already is), "life itself" might be for sale.

Which brings us to the other disjunction. Hayles' critical concern over how speculations (literary and otherwise) of a future time in which life is radically extended seem to always presume a parallel "immortality" for capitalism, and, more specifically of the kind of capitalist exchange that dominates the present, might be particularly suggestive here. As opposed to more ideal dreams of the future and of more progressive forms of social collectivity, descriptions of biological immortality seem to have traded one utopian version (a collective politico-economic one) for another (a potentially selfish "biological" one). The danger of such visions is that it seems to foreclose, in a sense, on the "future" itself, at least the vintage of the future that held some kind of liberatory power about fundamental changes to the more abject and self-interested aspects of contemporary humans. In this sense, then, it is no wonder that biological immortality would be so abhorrent to contemporary ethical thought as well as Platonically inflected turns to ontology within ethical thinking: visions of biological immortality seem to reassert the primacy of "the body"—not just the "actual" material body, but the selfish desires and appetites associated with the body in Platonic thought—as well as capitalist commodification—those very sources that Plato struggled so valiantly against in creating the foundation for Western ethical reason around the ethical subject and a form of valuation that might counter that of capital.

COMMODIFYING ETHICS; OR, DO THESE SHOES MAKE ME LOOK ETHICAL?

> All human virtue in circulation is small change; it is a child who takes it for real gold [*ächtes Gold*]. Nevertheless, it is better to small change in circulation than nothing at all. In the end, they can be changed into genuine gold [*baares Geld*], though at a considerable discount.
>
> —Immanuel Kant, *Anthropology from a Pragmatic Point of View* (44)

Two of the discourses I have discussed thus far—the Platonic invention of the immortal soul and the possibilities of concern for oneself manifesting as concern for others—were in many ways brought together in Foucault's final work on what he coded "the aesthetics of existence" or "technologies of the self." Theses processes, tracked by Foucault as crucial in the development of early Greek and Christian subjectivities, created a link between interiority or exteriority that could "permit individuals to effect by their own means, or with the help of others, a certain number of operations on their own bodies and souls, thoughts, conduct, and way of being, so as to transform them-selves in order to obtain a certain state of happiness, purity, wisdom, per-fection, or immortality" ("Technologies" 225). Such practices, for Foucault, were an additional category to the techniques of domination, truth games, and sign systems that occupied his earlier work, one in which an individual's attempts to master or control their own selves and bodies becomes ethical in its own right and leads to an ethical concern for others; one through which, as Foucault writes, a self that in thinking of itself, thinks of others. Foucault consistently insisted that the problems of thinking through this category of ethics was the same as that of the present, when "most of us longer believe that ethics is founded in religion, nor do we want a legal system to intervene in our moral, personal, private life" and "[r]ecent liberation movements suffer from the fact" that "they need an ethics, but they cannot find any other eth-ics than an ethics founded on so-called scientific knowledge of what the self is, what desire is, what the unconscious is, and so on" ("Genealogy" 256). For this is reason it is rather disappointing that his project on the genealogy of ethics ended, by choice or by chance due to Foucault's untimely death, in a rather familiar place. In Foucault's final seminars posthumously published as *Fearless Speech* and *The Courage of Truth*, self-sacrifice returns to ethical prominence in the Greek figure of the practitioner of *parrhesia* or "fearless speech," the individual who speaks truth to power while under the threat of personal or physical harm or even death, and who forms the opposite of the sophistic rhetorician.

Perhaps what is needed is a rethinking of technologies of the self based on transhuman excess rather than human finitude, a movement Deleuze antici-pated in the appendix to his book on Foucault in calling for the theorization of a "superfold" that emerges when traditional biology passes into "molecu-lar biology" and "the genetic code" and humans network with the "third generation machines" of cybernetics and information technology (131). Another way of stating this would be as a reimagining of the "aesthetics of existence" in a world dominated by what Virginia Postrel calls "the Aesthetic imperative," where self-fashioning and aesthetic enjoyment runs smoothly from the selection of infinitely customizable housing fixtures to the push for pharmacological and genetic manipulation of our appearances and bodies, and that emerges between a radical connectivity with others through global-ization and information technology as well as equally radical opportunities for self-indulgence through the relentless customization of experience.

Perhaps the best way to make this distinction is through an already-worn demarcation in cultural theory, and, more specifically, in the mani-fold historical relations between culture, aesthetics, technics, and ethics.

Specifically, I turn here to several representations of shoes, two of which have already been used to form a certain genealogy. First is a series of painting of shoes by Van Gogh, one of the subjects of Heidegger's "The Origin of the Work of Art," and later, partially through this initial reading, an important example in Jameson's hugely influential "Postmodernism" essay. In "Origin" Heidegger famously uses the image as a constellation point for thinking our relations to objects and commodities; although art's potentially revelatory powers may not lead to any transcendental or static truth, for Heidegger works such as Van Gough's painting have the potential to reveal what is often lost in a person's interaction with a real shoe: the connection between the human world of significant events and the nourishing or life-giving resources of the earth in the face of human mortality, what Heidegger calls a "shivering at the surrounding menace of death" (159). This is the quality of art that Heidegger saw disappearing in his present, a withdrawal that, as we saw in chapter two, he associates with the "end" of metaphysics and triumph of cybernetics.

Writing about a half-century after Heidegger, Jameson contrasts Van Gough's painting of shoes and this earlier interpretation with a series of

Figure 5.1 Van Gogh, *Shoes* (1886) (Van Gogh Museum, Vincent van Gogh Foundation, Amsterdam).

Figure 5.2   Andy Warhol, *Diamond Dust Shoes (Random)* (1980) (Image and Artwork ⌷ 2013 The Andy Warhol Foundation for the Visual Arts, Inc. / Licensed by ARS).

Andy Warhol paintings of "Diamond Dust Shoes," leveraging the presumed space between the cultural context and hermeneutic implications of the two to do a large amount of heavy lifting in marking the distance between an eclipsing "high modernism" and the emergence of postmodern culture and aesthetics. Although for Heidegger Van Gogh's painting might express a more authentic relation to the earth than that of the quotidian experience of modernity, and for Jameson it functions as at least a "Utopian gesture, an act of compensation" that might bring some positive return to capitalism's wholesale absorption of the human sensorium (7), Warhol's paintings, in which the representation of the shoes reproduce rather than resist their commodification, marks a gap in both hermeneutics and ethics; it frames the problem of ethical intervention in postmodern life, in which subjectivity is no longer simply alienated but increasingly fragmented.

All of this, I suppose, is rather familiar territory to readers of Heidegger and of Jameson's depiction of postmodernist culture as expressive of "a new kind of superficiality in the most literal sense" (9), a critique that has provided easy cannon fodder for conservative critiques of contemporary cultural "values" as well as spurred the more recent effort by left-oriented political theorists to search out deeper foundations via the return to ontology, meta-ethics, and retrofitted versions of various earlier philosophical conceptions of collectivity and the shared "being" of humans. However, despite the familiarity of these subsequent responses to the problematic posed within Jameson's reading, I want to hesitate a moment over the synecdochal power of these two examples before suggesting that we might need to add a third, one that is perhaps more timely than Warhol's and suggestive of a direction far different than the current return to ontology.

Jameson's central purpose in contrasting these two works is very much bound up with a certain consideration of "purpose" itself—that of the (perceived) object, of art, of the philosopher or critic, and the various appropriate and inappropriate uses that one might make of the other. While Jameson does not spend too much time on it, Heidegger's reading of Van Gough's work in "Origin" is also, of course, very much tied to these same questions. In considering the "peasant shoes" depicted in the painting, Heidegger imagines a hypothetical "peasant woman" that might own such shoes. For Heidegger, the shoes become a symbol of the "equipmental being of equipment," one of many examples he deploys in writings of this time to assay the degree to which humans can and commonly do access the "being" of objects and of themselves. Dwelling on the process through which his imagined peasant woman would trust in the "reliability" of the shoes, Heidegger suggests this relation functions at first as a way of securing oneself to the reliability or stability of the world, but eventually results in obscuring many other relations until only the "usefulness" of the shoes themselves is felt.

As this point, Heidegger suggests, the "worn-out usefulness of the equipment then obtrudes itself as the sole model of being, apparently peculiar to it exclusively. Only blank usefulness now remains visible" (160). If such shoes "themselves" stand in to symbolize the common, narrowed relationship between humans and objects and between humans, then the representation of the shoes in Van Gogh's painting, which deprives them of "their usefulness" as functioning technical objects, has the potential to model a more open and authentic series of relations for humans and more ethically cognizant and responsible forms of human sociality, collectivity, and connection to the earth. For Heidegger, then, Van Gough's work shows the *poesis* behind and beyond any instance of *techne*.[14] For Jameson, it is precisely this kind of power—the ability of the work of art to function in such a manner, really to have any hermeneutic purchase whatsoever—that seems impossible in Warhol's "Diamond Dust Shoes." Rather than presenting "a clue or symptom for some vaster reality," Warhol presents us with nothing more than "a random collection of dead objects hanging together on a canvas like so many turnips" (8).

Still further, we might dwell a moment on the how the two works are taken to reflect on the particular economic contexts shaping their creation and interpretation. If Heidegger understands the shoes immediately as "peasant shoes" and thus associates them with the sometimes brutal work of the agrarian laborer, by the time Jameson is contrasting the painting with "Diamond Dust Shoes," the relations of (aesthetic) mediation and capitalist commodification are taken as primary. Jameson feigns surprise that the work of Warhol, that former commercial illustrator whose entire work "turns centrally around commodification," fails to function as critical statements about that process and instead reveal "some more fundamental mutation both in the object world itself—now become a set of texts or simulacra—and in the disposition of the subject" (9). The simulacral quality of Warhol's "Diamond Dust Shoes" in particular undoubtedly comes from their imitation of the display of shoes that one would find in commercial art proper, as well as their positioning as a series of paintings that present variations on a single concept.[15] Most important for Jameson, however, is its exemplary power in explaining the "mutation in the object world itself," a phenomenon he will refer to elsewhere in the text as the reduction of culture around the question of *media*, a term that he takes as having conjoined "three relatively distinct signals: that of an artistic mode or specific form of aesthetic production; that of a specific technology, generally organized around a central apparatus or machine; and that, finally, of a social institution" (67). Here Jameson's relatively superficial and implicit consideration of the *poesis/techne* distinction in Heidegger's reading of the painting seems to return via an acknowledgment that the artwork is not so much unable to show the *poesis* behind *techne*, but how the two forces have become entirely indiscriminate and how their joining increasingly underlying social collectivity to form something we more commonly call "media." In this sense then, beyond the single question of the possible political or critical role of art in late twentieth-century culture, the function of the works as periodizing different regimes of ethics, economics, and politics *tout court* is emblazoned in their modeling of different forms of mediation, technicity, and sociality.

And it is with that expanded field of vectors in mind that I want to forward yet another object—again a representation, of sorts, of shoes—that might be a more relevant example, marking the not insignificant changes between the time of Jameson's writings and that of the present, what we might think of as a "further mutation" in the relationship between ethics, economics, and (aesthetic) mediation: the "(Product) RED Shoe." The (Product) RED Shoe is one of a variety of consumer goods created as part of the (Product) RED Campaign, a cooperative arrangement between a number of multinational corporations and the Global Fund to Fight AIDS, Tuberculosis and Malaria. Through this arrangement, a source of significant ad campaigns for many of the companies involved, a portion of all monies spent on often limited edition (RED)-branded versions of the company's typical products are collected and spent directly on the global fund's health initiatives. Though I have reproduced an image of one pair of the shoes here, it is fair to say that it is literally "unrepresentable" as a single object or image; rather, purchasers are invited to alter the color and design of the shoes at the point of purchase via a graphic interface. In this sense, we might take the shoe as an example of two processes I have been tracking over the course of this book: the parametric mode of technics, and the mode of algorithmic or open-source economics in which divisions between production and consumption as well between labor and leisure become intermixed. If Van Gough's painting of shoes was taken to encapsulate a certain spirit of labor in feudal capitalism given the aura of a work of art, and Warhol's "Diamond Dust Shoes" taken to signify the encroaching collapse between advertising and art proper as well as the becoming-commodity of images themselves, then perhaps the (Product) RED Shoe, all at once an individually manipulated image and physical commodity, is perhaps a more telling symbol of the forces currently dominating the fields that Jameson organized around the term "media" several decades ago.

More importantly, however, I take it that the (Product) Red Shoe is also a synecdoche for the contemporary form of the ethical and political questions that prompted Heidegger and Jameson to turn to these earlier works, particularly the connections between ethical capacity and regimes of commodification and communicative mediation, linkages already inextricable by as early as the time of Heidegger's writing. It would be easy to suggest, as many have, that purchasing such products is a poor substitute for actual political intervention. In fact, this is precisely how Tasmin Smith, the marketing head for (Product) RED describes the endeavor:

> We use the word "punk rock capitalism." There are some people who want to march on Washington or 10 Downing Street, and other people who just aren't that politically engaged and active. Red provides a very immediate empowering mechanism for someone to do something quite revolutionary, to cause a big corporation to break off a portion of its profit and put it towards a huge social challenge. (qtd. in Worth)

The director of the campaign, Sheila Roche, goes on to explain that (Product) Red is a kind of trendy, or feel-good altruism: "It's a way for the sinner

Figure 5.3   One of a multitude of pairs of customizable (Product) RED Converse shoes, in this case, the author's own.

to become saint—to spend money but to feel good about it. Red's hip and sex. Red is never about making a purchase because you're feeling sorry for someone" (qtd. in Worth).

However, I take it that what is unique about this approach to promoting charitable giving and social awareness, one modeled in several other similar campaigns of the last decade, is its odd mixture of altruism and cynicism.[16] On the one hand, the goal of at least the organizing branch of (Product) RED, if not necessarily the corporations that have become their partners, is undeniably invested in the fund- and awareness-raising goals of the endeavor. On the other hand, the operation seems designed to not combat but entirely work around the ostensibly passive nature of individuals, to make no concerted attempt to galvanize people around a health crisis that is at least partially attributable to massive differences in the economic resources between nations. Indeed, in addition to openly drawing on the consumer's preference to passive acts of purchase and collaboration with for-profit corporate entities, it is also worth noting that a variety of individuals purchasing these projects might be entirely indifferent, if not absolutely unaware, of the campaign and its involvement in aiding the medical treatment of impoverished individuals; given the large number of limited edition versions of popular products—most often the production of a commodity that is usually confined to a narrow color palette to be produced in the signature "red" of the effort—it seems certain that many are motivating solely by the desire to possess these "rare" editions. Still further, part of the pitch of the campaign, and presumably at least partially the motivation behind using distinct color schemes and insignia, is that many purchasers are guided by their desire to be able to display their altruistic spirit quite literally by wearing and carrying around their (Product) RED items. Thus, the manifold marketing strategy behind the campaign attempts to enlist individuals "authentically" invested in aiding in the altruistic work of the Global Fund, those that simply *want to be seen* as being the kind of person invested in that activity, as well as those *entirely oblivious* to the campaign and interested in the material product itself.

In this sense, then, the persuasive strategies encapsulated by the (Product) RED campaign might be better read not as a particular corruption of ethical relationships due to its fairly intense focus on the force of commodification, but rather as an appropriation of, or return to, the ethical strategies represented in the thought of the sophists and against which Platonic thought and ethics defined itself. Much like the sophistic counterpoints emphasized in the dialogues analyzed above, the Product (RED) campaign draws its power by taking the commodified relation as primary, as leveraging a certain "selfish selflessness" and the pressures of social *doxa*, and focuses its strategies on prioritizing "action" or performance that might be considered ethical over the creation of maintenance of ethical "subjects" or dispositions. The return of technics and rhetoric as a central force in shaping human sociality and communication may have brought with it a parallel return of the ethical environment in which these two domains were first defined. Indeed, though I took up the (Product) RED Shoes as a parallel object to Jameson's earlier examples, it seems such a "commodification of ethics" has been present, at least implicitly, in a wide variety of recent cultural phenomena:

—Certainly one might expand consideration of the dynamic outlined above in (Product) RED to a variety of changes in consumer preferences and marketing that do not have any specific attachment to a defined altruistic goal, but rather more amorphous connections to ethically and communally centered practices and economies. Perhaps best studied in this regard is the phenomenon in which consumers attach added value to goods and services—from "green" lawncare to hybrid and electric automobiles—that are less ecologically damaging than their counterparts. Here too it is important to note the often spectacular component of purchasing or using such goods; in a phenomenon some economists refer to as "conspicuous conservation" because of the ways it seems to invert the behavior that Veblen deemed "conspicuous consumption," it is not only the *performance* of eco-conscious consumption that is prized by purchasers but also its *appearance* in front of one's peers. If, in Veblen's schema, the leisure class's propensity to consume unnecessary commodities is seen as "honourable . . . as a mark of prowess and a perquisite" and eventually becomes "substantially honourable in itself" in the social consciousness (69), we have seen the reverse take place in the present: one demonstrates honor and ethical *bona fides* by making their consumption appear as less wasteful and more directed toward concern for the well-being of others. The recent upsurge in interest for handmade goods and "simple living," as well as locally sourced and "cruelty-free" foods, and so-called ethical consumerism of all stripes also speak to the ways in which consumer desire and commodity fetish drive contemporary "anti-corporate" sentiment just as much as they drive traditional corporate capitalism. However, the best example of the odd marriage of vulgar marketing technique with traditionally progressive ethical concern that marks the last few decades probably remains American Apparel; the manufacturer has succeeded in making it popular to purchase clothing produced through an environmentally conscious process, made by employees who receive a living wage, and, in several popular lines, advocating strongly for such progressive causes as gay marriage and the granting of amnesty for illegal immigrants in the U.S., but done so largely through some of the most consistently sexually exploitive marketing campaigns in the history of modern advertising.

—We might see this tendency in the changing shape of protest movements, notably the intersecting endeavors of the *indignados*, Occupy, and "We are the 99%" initiatives, all of which have adapted algorithmic strategies of contemporary capitalism and its concordant marketing logics. The novelty of these radical movements comes, on the one hand, in their default positioning of capitalism as in need of repair rather than replacement; as Michael E. Connolly has emphasized, while the Occupy movement certainly gives lie to "neoliberal economic fantasies" that the market can regulate itself with little government intervention, as far as one might attribute discernable demands for redress from participants, they are those that are eminently at home with the core qualities of capitalism: "production for profit, contractual labor, the primacy of the commodity form, a significant degree of competition between firms, and a large role for the state" ("What"). On the other hand, the combination of the wide range of grievances and experiences represented within the group and the presentation of such via forms of social media demonstrates an ingenious (re)appropriation of capitalist niche and viral marketing schemas. The thousands of searchable testimonials posted on the popular "We are the 99%" Tumblr blog, for instance, both attract and provide an eminently flexible series of occasions for empathy and recruitment. Through such strategies, these movements seem to have solved a challenge facing protest movements since at least the aftermath of the events of May '68 in Europe: that of balancing the ostensible minority status of such protests—necessary because of neglect from the mainstream media or the general communicative regime in which they are embedded—with their necessity of making use of the same techniques in order to "appear at all," to be noticed by other individuals within the same environment.

—The direct application of the mercenary logic of commodification to such questions as worker and consumer safety, equal access to health-care, environmental crises, and even international warfare has also been one of the more novel and notable strategies of progressive political legislation over the last few years. At first blush, the identification of a human life to a discrete dollar value would seem to be the worst example of capitalist excess and its dehumanizing tendencies. However, to give two examples, the 2010 decisions by the U.S. Environmental Production Agency and Food and Drug Administration to set the value of a human life at $9.1 and $7.9 million, respectively, was more immediately controversial for the pressures it set on corporations and subsidiary government agencies to protect the health and safety of both their employees and individuals who receive their products and services.[17] The commodification of the "right" to pollute the atmosphere via emissions trading and carbon exchange schemes inside and across national boundaries has similarly been effective in slowly reducing emissions dangerous to ecological sustainability, and in promoting international cooperation on environment protection.[18] Finally, support for everything from military withdrawal from the Iraq War to the passage of partially nationalized healthcare in the U.S. has been forwarded along more successfully via fairly rigid "cost-benefit" analyses calculated in fiscal rather than moral terms, making transversal connections between appeals to economic reason and those of social justice.

—In closer relation to the intersection of populist politics and ethics, we might also consider the strategies of so-called "triangulation" or (more positively) "radical center" strategies that became particularly prominent in political discourse of the U.S. and Europe starting in 1990s. Here too we might suggest that the aspects of this phenomenon that have received the most condemnation—the ideological inconsistency or flexibility of its proponents, the active appropriation of agendas traditionally associated with one's political opponents, from Clinton's famous announcement that "era of big Government is over" to Bush's embrace of social justice issues of a type via a "compassionate conservatism"—are the same ones that at least show the potential for a novel model of political engagement and mobilization in regards to ethical issues. Take, for instance, one of Clinton's more oft-cited acts of "triangulating" a divisive issue: his pronouncement at the 1996

Democratic National Convention that abortions should be "safe," "legal," and "rare." Clinton's suggestions that he may hold a moral objection to abortion while insisting that he would defend its legality was frequently taken, at worst, to be an overwhelmingly hypocritical or opportunistic move, and, at best, an at least potentially laudatory commitment to maintaining morality and legality as separate domains.[19] What is perhaps most notable about such a strategy, however, is its denigration of commitment to particular ethical subject positions or ideological investments in favor of a pure focus on the desired "effects" of such positions, one that additionally demarcates rhetoric, or persuasion, as a field that itself might be considered distinct from morality and legality. Attacks on the authenticity or consistency of politicians engaged in such strategies, on the other hand, start to look more like moribund, or even oddly quaint, presumptions of the importance of such qualities in the face of a becoming-*techne* of ethics and politics.

—A similarly strange appropriation of the logic of capital in service of goals more commonly associated with socialist or anti-capitalist aims can be found in the recent upsurge of local exchange trading systems (LETS), complimentary and local currency, time banks, and other forms of collaborative financing. Largely driven through motivations that seem to mix historical memory for a variety of eighteenth- and nineteenth-century anarchist and "associationist" movements that were largely crowded by Marxism and socialism, as well as the more general obsession for "localism" and prosumption consumer trends, alternative economies of these types have surged internationally following the 2008 financial crisis, sprouting up everywhere from Medellín to Brixton to Detroit.[20] Though similarly demonstrative of the "corrective" rather than "resistant" approach to institutional capitalism seen in direct social protest after the financial crisis, these economic collectives are perhaps most notable for the ways they generally invert the traditional positioning of production and consumption within (post-)Marxist theory: participants' ties of solidarity within such communities, as well as their agency in impacting the larger capitalist economies within such systems, reside in their role as consumers of goods and services and exchangers of currency, rather than as workers or producers.

What the above examples have in common is how they demonstrate the ways in which, in at least partial response to the rise of parametric media and the increasing *technicity* of the socius, the forces and phenomena traditionally coded as ethics in the Western culture—social collectivity, moral decision making, altruism, self-understanding—have become imbricated with the logic of capitalist commodification. In this sense we might say that we have now come full circle from the concerns with which we began this chapter. If ethics emerged as a response to the formalization of commodity

capitalism under the *ne plus ultra* commodity form of coined money, bringing with it the immortal soul as "moral subject" and the cultural triumph
of philosophical reason over sophistic *techne*, then the last few decades
seem to have tracked a return of the repressed if not an entire inversion of
this earlier process: at a moment during which *techne* has returned to the
forefront of contemporary life, and one in which immortality, if not all the
joys and horrors of heaven and hell, now seem realizable in the material
world, we have seen the return of monetary abstraction as a ruling force in
the contemporary possibilities for ethics.

All of this, of course, seems to leave us with a series of pressing questions. Can we understand the "commodification of ethics" as something
more than simply a particularly egregious example of the increasing imbrication of economics and culture typically bemoaned by critics of capitalist
encroachment into the social realm? Is there a way to distinguish the strategies and practices listed below—in which the logic of capitalist exchange
and marketing is appropriated for ameliorative means—from the more
familiar, and certainly more problematic, conflation of capitalism *as an*
ethical system or ideology, a suggestion that might be traced as far back
as Aristotle but is now more commonly associated with the "neoliberal
moralism" of thinkers like Friedrich Hayek and Ayn Rand? In short, is this
intrusion of the logic of commodification into the realm of ethics nothing
more than a symptom of the triumph of capitalism itself?

On another read, however, we might find two crucial problems with
these questions. First, they too easily mistake a certain mode of abstraction—that of "commodification" through which value accrues to something via a socialized mode of exchange—and a formal economic system.
Furthermore, we might suggest, as we have earlier in this text, that it is
immensely difficult to define or discriminate capitalism "itself," as opposed
to, for instance, particularly positive or negative tendencies often associated with it, from both other forms of economy or contemporary sociality
as a whole. Second, we might ask after whether the specific allergy to the
conflation of the logic of commodification and ethics additionally relies on
an even more problematic presumption that one can position some kind
of "outside" to the flows of desire and attachment created and circulated
through such exchange, an interior space free of or resistant to its influence,
a role traditionally play by the "soul" or in more recent times thematized
ontologically as the authentic or ethical subject. Perhaps more pointedly,
we might wonder why, after decades of studies devoted to the "social construction" or performativity of human identity, we continue to repeat the
Platonic insistence that "good actions" must be the result of "good actors"
and orient our ethical judgment and decision making around this distinction. Finally and most pragmatically, however, the very occurrence of the
moment in which the logic of commodification becomes inextricable from
that of the socius would seem to suggest that one cannot be really *for* or
against it. The becoming-*techne* of social relations of all stripes that has

intensified over the last several decades has perhaps only made apparent a secret the sophists revealed a long time ago: that "moral value" is just one among many others, something to be put to work rather than to be attained as its own end.

In other words, as a particularly intense cultural response to the *technicity* of contemporary social life, the commodification of ethics is not so much something that one can judge, but something everyone must respond to or work through in order to see how it might be used to forward particular ends, or be deployed in ways counter to its more dangerous or destructive purposes. In this sense, the return to more traditionally metaphysical notions of the subject, as well as the return of overdetermined community identities and cultural fundamentalisms of various kinds might be taken as itself a regressive measure: an effort to find a space outside or "counter" to the seemingly ubiquitous *technicity* of the present as revealed in the commodification of the social.

If there is any unifying thread subtending the various changes in technoscience, media forms, and political economy that I have been tracing under the rubric of the Cybernetic Age throughout this book, it is that the increasing *technicity* of everyday life has made it increasingly impossible, or one might say *too costly*, to orient our hopes for ethical and political change on the economies of the natural, the meaningful, or the authentic and ideal, nor can one presume any advantage to the generic rejection or challenging of these same categories. Rather, an increasingly essential component of everyday struggles of all types will be fought through the always already compromised forces of rhetoric, those of persuasion, technique, and the creation and manipulation of desire. Insofar as these are the forces that structure contemporary social field, they at the same time provide our best options for the political and ethical challenges of the present. And what of the future, that concern with which we began this chapter? We might say that our relationship to it is the one thing that has not changed: it so rarely turns out to be what we hoped, but almost always to be what we desired.

Notes

NOTES TO THE INTRODUCTION

1. For a concise review of Ampere's selection of this term, and a partial translation of Ampere's description of its coinage, see Dyson (5–7).
2. For a remarkable history of the origins of the cybernetics group and the Macy Conference meetings, see Heims.
3. See Beers for the history of Project Cyberin.
4. A compelling argument for the significant and wide-ranging influence of cybernetics on mid-century poststructuralist thought can be found in Lafontaine.
5. For an overview of this position and its detractors, see Schiappa.
6. Despite what might seem to be a fairly intuitive connection between rhetoric as a series of feedback-based techniques of persuasion and the variety of forces associated with contemporary technology and new media that also seem to emphasize flexibly-responsive forms of communication and interaction, there has been a surprisingly small amount of work on these similarities. New media scholar and video game designer Ian Bogost, for instance, has recently suggested that procedural rhetoric seems to be a primary technique used in structuring our interactions with video games and software more generally. However, perhaps more common is the perspective suggested by Lev Manovich in his highly influential *The Language of New Media*, that rhetoric seems to be on the "decline" in post-literate culture because of our tendency to associate it with oral communication, and later, traditional print media (77). For notable exceptions to this trend within rhetorical scholarship, see, in particular, Brooke and Welch for two works that explicitly retrofit ancient rhetorical principles to new media.

NOTES TO CHAPTER 1

1. See Henderson for details.
2. Reported in, among other places, Meek.
3. For coverage see Weiss.
4. The group's preliminary findings were subsequently published in 2005 by the National Research Council as part of the volume *Guidelines for Embryonic Stem Cell Research*.
5. For coverage of Woolley's death, see Miller.
6. Although media accusations that video games are addictive and/or encourage violent behavior are numerous, the Woolley case reflects recent

technoscientific research into the neurology of video gaming. See Koepp for the relationship between video gaming and dopamine.

7. See the coverage in Cass.

8. See also, for instance, Penn's descripton of "psychographic microtargeting" in Clinton's 2006 campaign (145–146). For an overview of microtargeting written during the run-up to the 2008 presidential election, see Svoboda.

9. See, for instance, *Discipline* (181). "Capacitation," read as a "rendering capable," also has the advantage of having extant implications within a variety of discourse that we will encounter below, such as denoting potential action in the life sciences (particularly in relation to reproduction), machine technologies (such as in electronic capacitors), and legal discourse (to qualify or legitimate an individual or a law).

10. See, for example, *Critique* (387–390).

11. Interestingly enough, as Horgan does not fail to mention early in *The End of Science*, Gunther Stent was also one of the scientists who participated in the conference at Gustavus Adolphus College that is behind the similarly named collection also under review here.

12. See CBC News for text from Goodyear's proposal.

13. See, in particular, Marx's planned "seventh part" of the first volume of *Capital*, "Results of the Immediate Process of Production," included as an appendix in *Capital* (941–1084).

14. Antonio Ceraso and I have developed this idea more generally around the question of "open source" as a production method; see Ceraso and Pruchnic.

15. For a concise history of neuromarketing and its attendant controversies, see Fisher et al.; for a recent overview by a practitioner, see Lindstrom.

16. For the critical contrast between affect and ideology, in addition to Massumi and Sedgwick, see, for instance, Grossberg (79–108).

17. See Simon, and Marcus, Neuman, and MacKuen.

18. There is much in the history of both contemporary computer science and contemporary affect theory to support such a conclusion. On the one hand, Silvan Tompkins immensely influential work on human affect, as Sedgwick herself notes, emerged from his initial interests in the possibility of simulating human qualities in mechanical realms (see Sedgwick and Frank). On the other side of the equation, much prototypical work in computer science was based around the simulation of human nervous and neural systems activity (see, e.g., Anderson and Rosenfeld).

NOTES TO CHAPTER 2

1. See Derrida, "Letter" (270–271), and Heidegger, *Being* (41–49).

2. The comparison below was greatly informed by Crosby (12–19).

3. For instance, according to historian John North, the announcement of such items as curfew times or the close of the public market via the chimes of the famous fourteenth-century clock in the abbey of St. Albans took the place of earlier signals produced manually by bell-wringers (219–220).

4. Deleuze seems to be suggesting something of this idea in stating in that "in Baroque the coupling of material-force is what replaces matter and form" (*The Fold* 29).

5. To give just two more examples, computer scientist Joseph Weizenbaum, whose invention, the the ELIZA natural-language process system, inspired more than a few outsized claims for the viabilty of artificial intelligence in computing environments, subtitled his 1976 critique of this trend "From

Judgment to Calculation"; Ian Hacking has emphasized the risk that we may "replace judgment by computation" in overemphasizing the supposed objectivitiy of policy and military actions supported by "decision theory couched in terms of probability" (4).

6. For the Greek case, see, in particular, Nightingale.
7. See de Man (*Blindness*); Zizek (*Parallax*); Lyotard, "The Return"; and Levinas, *Totality* (121–142).
8. As architect Mutsoro Sasaki suggests, referring to his own use of computer-aide design, such techniques allow one "to create unknown but logical structural forms beyond our empirical knowledge" (68–69).
9. Rem Koolhaas, for instance, has remarked on how the Chinese Central Television headquarters building (2002) he helped designed would have been impossible to realize ten years ago, not because of advances in structural materials or design principles, but because of "the sheer time of computing required" to generate the structure's design (qtd. in Anderson).
10. See Habermas, *Structural* (168–175).
11. See for instance Braudel's survey of "early" world economies (71–88).
12. For early controversies over futures markets see Levy, and Mulherin et al. (620–624).
13. To give just one quick, but important example, consider Marx, surely one of the prime contenders for "first" or prototypical critical theorist. Although we might tend to think of Marx's treatment of ideology as one in which the ideological covering of material life is subjected to critique and thus eliminated, as Althusser emphasizes, for Marx ideology was a necessity for human existence, and this necessity demanded that politically-minded theorists "act on ideology and transform ideology into an instrument of deliberate action on history" (232).

NOTES TO CHAPTER 3

1. Burke's description of the "unending conversation" can be found in *The Philosophy of Literary Form* (110–112).
2. See Burke, "Order," for his discussion of kind/degree differences between humans and animals (171).
3. For an account of first-wave cyberneticists' response and resistance to behaviorism, see Heims (1–13, 201–247).
4. Burke comments on both of these impulses in "Counter-Gridlock": "That's where the deconstructionist guys are cutting in, on that sort of thing. I want to stay halfway there. Destroy it, yes, if you will. But first let us see it as having the form it does, with its particular kind of beginning, middle, and end" (22).
5. Of course, Burke does provide prescription in the "Program" essay of *Counter-Statement* (although he would later advise that this essay is effective only if not taken "literally"). Hicks famously condemned this aspect of the book as well, arguing that "it merely describes the social attitudes of a man who is principally interested in technique."
6. Burke outlines his hypothesis concerning the decline of formal interest in rhetoric in the conclusion to his "Lexicon Rhetorica" (*Counter-Statement*, 210–213), attributing the "revolt" against rhetoric as the unfortunate side-effect of a more specific resistance to the ceremonious, and associating the return of rhetoric with a renewed interest in reconsidering aesthetics and persuasion as discourses aimed at producing particular effects rather than transmitting knowledge or information.

7. Burke quotes from both an 1852 letter to Louise Colet (*The Letters of Gustave Flaubert, 1830–1857*, 154–155) and an 1876 letter to George Sand (*The Letters of Gustave Flaubert, 1857–1880*, 232–233).
8. One could, of course, productively contrast Massumi's satirical take on this reductive definition with Burke's own exhaustively extended "Definition of Man."
9. See Worsham for nuanced accounts of this conception.
10. See Ballif (153–194) and Davis (21–115) for examples of the former movement.
11. Burke refers to Bergson and Nietzsche as practitioners of "perspectivism" in the afterword to *Permanence and Change* (311).

NOTES TO CHAPTER 4

1. The duo's press releases were previously available at the now inactive home-lessweek.com site.
2. Weeks' former site, droppingbombsonyourmom.com, is currently inactive.
3. See also Galloway for a similar analysis that uses the network configuations of the Internet as a synecdoche for power in network society.
4. Of course, the homeless can—and certainly have been—disciplined in a Fou-cauldian sense. For example, a large portion of Mike Davis' *City of Quartz* is devoted to detailing ways in which the homeless in Los Angeles have been corralled to disparate sites of confinement through various attraction and repulsion mechanisms.
5. This is not to suggest that the homeless and poor do not participate in capital-ism as consumers. As Deleuze and Guattari (drawing on the work of Baran and Sweezy) explain in *Anti-Oedipus*, "What on the contrary is called the co-opting power of capitalism can be explained by the fact that its axiomatic is not more flexible, but wider and more englobing. In such a system no one escapes participation in the activity of antiproduction that drives the entire productive system" (236). Though the impoverished may not be considered "good" consumers in certain economic formulations—as their power of choice is certainly diminished—they certainly at the very least participate in antiproduction through the benefices of various governmental and non-governmental agencies.
6. Deleuze writes that "Burroughs was the first to address" this new formation ("Control" 174) and that "Control" was the title "proposed by Burroughs to characterize the new monster" that is control society ("Postscript" 178).
7. Here I refer to Deleuze's oft-noted tendency to produce creative interpreta-tions of other canonical thinkers works as seen, for instance, in his extended studies of such figures as Nietzsche, Foucault, Spinoza, and Bergson.
8. Foucault makes his critique of ideology as a critical concept clear in the inter-view "Truth and Power," highlighting the "longing for a form of power inno-cent of all coercion, discipline, and normalization" he finds to be implicit in its use in analyzing culture (117).
9. One of the most influential, Foucault-inspired strategies of this kind is, of course, Judith Butler's suggestion of methods for "subverting" gender identi-ties in *Gender Trouble* (see, in particular 171–190).
10. See, for instance, Barthes, "Rhetoric of the Image," and Williams, "Advertis-ing: The Magic System."
11. Consider, for instance a contemporary marketing text chosen more or less at random: Andreas Buchholz and Wolfram Wördemann's *What Makes Win-ing Brands Different?* This book contains a chapter titled "Perceptions &

Programs," a title that of course would not be out of place in *What Is Philosophy?* One of the five strategies outlined is this chapters is "Inverting a Negative," which asks the reader to consider, "Are the weaknesses consumers perceive in your brand really weaknesses? Are the strengths they perceive in your competitor's product really strengths? You can turn a perceive liability of your brand into an asset by changing the consumer's perception. The converse is true when it comes to effectively fending off competitors" (76). This is, of course, the sophistic rhetorical tactic of "making the weaker argument the stronger."

12. Agamben, "Bartleby, or On Contingency," *The Coming Community* (37), *Homo Sacer* (48), and *Idea of Prose* (65, 78); Blanchot, *The Writing of the Disaster* (17); Derrida, *The Gift of Death* (65); Hardt and Negri, *Empire* (203–204); and Zizek, "Notes" and *The Parallax View* (375–383). Deleuze's own reading of "Bartleby" can be found in Deleuze, "Bartleby; or, The Formula."

NOTES TO CHAPTER 5

1. See Cohan and Sweet for estimates.
2. See also Steiner and Kurke.
3. For examples of the recuperation of the sophists as neo-relativists or "proto-postmodernists," see, for instance, Vitanza (particularly 27–55), Ballif, and McComiskey. For examples of the relatively rarer works that take up the Sophists as moral philosophers, see Bett and McKirahan (353–389)
4. Indeed, Nietzsche would echo more or less all of Callicles' critiques, arguing in "Homer's Contest" that Socrates' indulgence in rhetoric against the sophists is primarily done in order to proclaim superiority over them, "his finally being able to say 'Look: I, too can do what my great rivals do; yes, I can do it better than them'" (179); Nietzsche will also indict the "disrespect and superiority" implicit in Socrates' condemnation of the popular in *The Birth of Tragedy* (66).
5. Again, Badiou is the exception, given his explicit commitment to Plato's views on the role of philosophy and the importance of pursuing "truth" and "thought" as categories (see, e.g., Badiou and Hallward 119–120).
6. Recall, for instance, how in one of the "founding texts" of critical theory, Adorno gets the endeavor off of the ground via a rejection of "that question which today is called radical and which is really the least radical of all: the question of being *(Sein)* itself " ("Actuality" 121).
7. These examples found, among other places, in Zizek, *Sublime* (116–118), *Fragile* (151–160), and *Ticklish* (382–292).
8. Derrida's formal consideration of Marx's theories of commodification and the relation between commodity fetishism and religious fetishism and messianism in *Spectres of Marx* also in many ways forms the segue to his late work on ethics and justice. See *Spectres* (186–210).
9. As Strathausen writes, those taking part in what he calls neo-left ontology share an emphasis on imagining the "historically contingent construction of a different 'nature' from the one we presently inhabit" (19).
10. For Derrida on "messianic time" see *Spectres* (61–95). For the "future to come" *(l'a-venir)* see, for instance, *Archive Fever* (68). For "democracy to come," *Politics* (232–233).
11. Despite his subsequent association with a creative design think-tank and other religiously-inclined organizations, in *The Physics of Immortality*, Tipler suggests he should be best considered "an atheist" at present because

his work suggests only that a something fulfilling the criteria of "God" does not yet exist and his predictions that something like this will occur have not been definitively proven (305–306).

12. Notable in this regard is the work of Robert Ettinger, who as the "inventor" of cryonics could be claimed to be the first in such a series. His mid-'60s and mid-'70s works *The Prospect of Immortality* and *Man Into Superman* show a consistent interest in questions of ethics and sociopolitical crises. For a more contemporary iteration see, in addition to Kurzweil, Aubrey de Grey's "Life Extension, Human Rights, and the Rational Refinement of Repugnance."

13. For a particularly prominent example of the former, see Agamben, *Homo Sacer*, for the latter, see Hardt and Negri, *Empire* (22–41).

14. Or, as Nealon states it more directly, what Heidegger seems to be arguing in this case is that the "essence of *techne*, in other words, is *poesis*" (*Alterity* 98).

15. Nealon, in writing about the general problem of the "examplary" in (post) postmodern culture has emphasized the strangeness of Jameson failing to mention that the Van Gogh painting referred to by Heidegger is itself part of a larger series of similar works by the artist (*Alterity* 100–109).

16. To give one more example, the Unicef Tap Project® is a campaign in which participating restaurants sell a glass of (usually complimentary) tap water to customer for $1; this water is typically served in special glass that advertises the patron's contribution and the money is devoted to improving international access to clean water by impoverished communities.

17. For coverage, see Applebaum.

18. See A. Denny Ellerman's study *Pricing Carbon*.

19. For an example of the latter view, see Rasch (146–147).

20. The Complimentary Currency Resource Center currently houses a database of worldwide local exchange trading systems on their website.

References

Adorno, Theodor W. "The Actuality of Philosophy." *Telos* 31 (Spring 1977): 121–133. Print.

Agamben, Giorgio. "Bartelby, Or On Contingency." *Potentialities: Collected Essays in Philosophy*. Ed. and Trans. Daniel Heller-Roazen. Stanford, CA: Stanford UP, 1999. 243–272. Print.

———. *The Coming Community*. Trans. Michael Hardt. Minneapolis: U of Minnesota P, 1993. Print.

———. *Homo Sacer: Sovereign Power and Bare Life*. Trans. Daniel Heller-Roazen. Stanford, CA: Stanford UP, 1998. Print.

———. *Idea of Prose*. Trans. Michael Sullivan and Sam Whitsitt. Albany: State U of New York P, 1995. Print.

———. *The Open: Man and Animal*. Trans. Devin Attell. Stanford, CA: Stanford UP, 2004. Print.

Althusser, Louis. "Marxism and Humanism." Trans. Ben Brewster. *For Marx*. New York: Verso, 2005. 219–247. Print.

Anderson, James A., and Edward Rosenfeld. *Talking Nets: An Oral History of Neural Networks*. Cambridge, MA: MIT P, 1998. Print.

Anderson, Kurt. "From Mao to Wow!" *Vanity Fair*. Web. 19 Nov. 2010.

Ansen, Alan. "Anyone Who Can Pick Up a Frying Pan Owns Death." *Big Table* 2 (Summer 1959): 32–41. Print.

Applebaum, Binyamen. "As U.S. Agencies Put More Value on a Life, Businesses Fret." *New York Times*. Web. 16 Feb. 2011.

Arendt, Hannah. *The Human Condition*. 2nd ed. Chicago: U of Chicago P, 1998. Print.

Aristotle. *Mechanics*. Trans. E. S. Forster. *The Complete Works of Aristotle*. Vol. 2. Ed. Jonathan Barnes. Princeton, NJ: Princeton UP, 1984. 1299–1318. Print.

———. *Rhetoric*. Trans. W. Rhys Roberts. In *The Complete Works of Aristotle*. Vol. 2. 2152–2269. Print.

Arrighi, Giovanni. *The Long Twentieth Century: Money, Power, and the Origins of Our Times*. New York: Verso, 1994. Print.

Badiou, Alain. *Ethics: An Essay on the Understanding of Evil*. Trans. Peter Hallward. New York: Verso, 2001. Print.

Ballif, Michelle. *Seduction, Sophistry, and the Woman with the Rhetorical Figure*. Carbondale: Southern Illinois UP, 2001. Print.

Bargh, John A., and Tanya L. Chartrand. "The Unbearable Automaticity of Being." *American Psychologist* 54.7 (1999): 462–479. Print.

Barthes, Roland. "The Old Rhetoric: An *aide-mémoire*." *The Semiotic Challenge*. Trans. Richard Howard. New York: Hill and Wang, 1988. 11–94. Print.

———. "Rhetoric of the Image." *Image, Music, Text*. Trans. Stephen Heath. New York: Hill and Wang, 1977. 32–51. Print.

Bateson, Gregory. "The Cybernetics of 'Self': A Theory of Alcoholism." *Steps to an Ecology of Mind: Collected Essays in Anthropology, Psychiatry, Evolution, and Epistemology.* Northvale, NJ: Jason Aronson, 1987. 309–337. Print.

Bauman, Zygmunt. "Is There a Postmodern Sociology?" *Theory Culture Society 5* (1988): 217–237. Print.

Beer, Stafford. *Platform for Change.* New York: Wiley, 1975. Print.

Beller, Jonathan. *The Cinematic Mode of Production: Attention Economy and the Society of the Spectacle.* Hanover, NH: Dartmouth College P, 2006. Print.

Berardi, Franco "Bifo." *After the Future.* Trans. Arianna Bove, Melinda Cooper, Erik Empson, Enrico Giuseppina Mecchia, and Tiziana Terranova. Ed. Gary Genosko and Nicolas Thoburn. Oakland, CA: AK Press, 2011. Print.

Bergson, Henri. *Laughter: An Essay on the Meaning of the Comic.* Trans. Cloudesley Brererton and Fred Rothwell. Los Angeles: Green Integer, 1991. Print.

———. *Matter and Memory.* Trans. Nancy Margaret Paul and W. Scott Palmer. New York: Zone, 1991. Print.

———. *Time and Free Will: An Essay on the Immediate Data of Consciousness.* Trans. F. L. Pogson. Mineola, NY: Dover, 2001. Print.

Berlant, Lauren. "The Subject of True Feeling: Pain, Privacy, and Politics." *Cultural Pluralism, Identity Politics, and the Law.* Ed. Austin Sarat and Thomas R. Kearns. Ann Arbor: U of Michigan P, 1999. 49–84. Print.

———. "Unfeeling Kerry." *Theory & Event* 8.2 (2005). Web. 12 Nov. 2012.

Berman, Marshall. *All That Is Solid Melts Into Air: The Experience of Modernity.* 2nd ed. New York: Penguin, 1988. Print.

Bett, Richard. "Is there a Sophistic Ethics?" *Ancient Philosophy* 22.2 (2002): 235–262.

Blanchot, Maurice. *The Writing of the Disaster.* Trans. Ann Smock. Lincoln: U of Nebraska P, 1995. Print.

Bogost, Ian. *Persuasive Games: The Expressive Power of Video Games.* Cambridge, MA: MIT P, 2010.

Braudel, Fernand. *The Perspective of the World. Civilization and Capitalism 15th–18th Century.* Vol. 3. Trans. Siân Reynolds. New York: Harper & Row, 1982. Print.

Brooke, Collin Gifford. *Lingua Fracta: Toward a Rhetoric of New Media.* Cresskill, NJ: Hampton, 2009. Print.

Brown, Wendy. "Untimeliness and Punctuality: Critical Theory in Dark Times." *Edgework: Critical Essays on Knowledge and Politics.* Princeton, NJ: Princeton UP, 2005. 1–16. Print.

Buchholz, Andreas and Wolfram Wördemann. *What Makes Winning Brands Different? The Hidden Methods Behind the World's Most Successful Brands.* Chichester, NY: Wiley, 2000. Print.

Buenza, David, and David Stark. "Tools of the Trade: The Socio-Technology of Arbitrage in a Wall Street Trading Room." *Industrial and Corporate Change* 13.2 (April 2004): 368–400. Print.

Burke, Kenneth. *Attitudes towards History.* Berkeley: U of California P, 1984. Print.

———. "Biology, Psychology, Words." *Dramatism and Development.* Barre, MA: Clark UP, 1972. 11–32. Print.

———. "Counter-Gridlock: An Interview with Kenneth Burke." *All Area* 2 (1983): 6–33. Print.

———. *Counter-Statement.* Berkeley: U of California P, 1968. Print.

———. "Definition of Man." *Language as Symbolic Action: Essays on Life, Literature and Method.* Berkeley: U of California P, 1968. 3–24. Print.

———. *A Grammar of Motives.* Los Angeles: U of California P, 1969. Print.

———. "Mind, Body, and the Unconscious." In *Language as Symbolic Action.* 63–80. Print.

———. "Order, Action, Victimage." *The Concept of Order.* Ed. Paul G. Kuntz. Seattle: U of

Washington P, 1968. 167–190. Print.

———. *The Philosophy of Literary Form: Studies in Symbolic Action.* 3rd ed. Berkeley: U of California P, 1973. Print.

———. *The Rhetoric of Religion: Studies in Logology.* Berkeley: U of California P, 1970. Print.

Burke, Kenneth, and Granville Hicks. "Counterblasts on *Counter-Statement.*" *The New Republic* 9 (1931): 101. Print.

Burroughs, William S. "Academy 23." *The Job: Interviews with William S. Burroughs by Daniel Odier.* New York: Penguin, 1989. 123–224. Print.

———. *Cities of the Red Night.* New York: Picador, 2001. Print.

———. *Last Words: The Final Journals of William S. Burroughs, November 1996–July 1997.* Ed. James Grauerholz. New York: Grove, 2000. Print.

———. "The Limits of Control." *The Adding Machine: Selected Essays.* New York: Arcade, 1993. 167–170. Print.

———. *Naked Lunch.* The Restored Text. Ed. James Grauerholz and Barry Miles. New York: Grove, 2001. Print.

———. *Nova Express.* New York: Grove Press, 1987. Print.

———. "The Revised Boy Scout Manual." *RE/Search* 4/5 (1982): 5–8. Print.

Burroughs, William S., and Brion Gysin. *The Third Mind.* New York: Viking, 1978. Print.

Butler, Judith. *Gender Trouble: Feminism and the Subversion of Identity.* 10th Anniversary Edition. New York: Routledge, 1999. Print.

———. *Subjects of Desire: Hegelian Reflections in Twentieth-Century France.* New York: Columbia UP, 1999. Print.

Carpo, Mario. *The Alphabet and the Algorithm.* Cambridge, MA: MIT P, 2011. Print.

Cass, Connie. "Addiction to Porn Destroying Lives, Senate Told: Experts Compare Effect on Brain to that of Heroin or Crack Cocaine." *Associated Press.* Web. 18 Nov. 2004.

Castells, Manuel. *The Rise of the Network Society.* 2nd ed. *The Information Age.* Vol. 1. Malden, MA: Blackwell, 2000. Print.

CBC News. "National Research Council to 'Refocus' to Serve Business." Web. 6 March 2012.

Ceraso, Antonio, and Jeff Pruchnic. "Open Source Culture and Aesthetics." *Criticism* 53.5 (Summer 2011): 337–375. Print.

Cohan, Peter. "Big Risk: $1.2 Quadrillion Derivatives Market Dwarfs World GDP." *Daily Finance.* Web. 6 June 2010.

Connolly, William E. "What Is to Be Done?." *Theory & Event* 14.4 (2011 Supplement). Web. 12 March 2012.

Cornford, F. M. *From Religion to Philosophy: A Study in the Origins of Western Speculation.* Mineola, NY: Dover, 2004. Print.

Crable, Brian. "Symbolizing Motion: Burke's Dialectic and Rhetoric of the Body." *Rhetoric Review* 22.2 (2003): 121–137. Print.

Crusius, Timothy W. *Kenneth Burke and the Conversation after Philosophy.* Carbondale: Southern Illinois UP, 1999. Print.

Crombie, A. C. "Quantification in Medieval Physics." *Isis* 52.2 (1961): 143–160. Print.

Crosby, Alfred W. *The Measure of Reality: Quantification and Western Society, 1250–1600.* Cambridge: Cambridge UP, 1997. Print.

Davis, D. Diane. *Breaking Up [at] Totality: A Rhetoric of Laughter.* Carbondale: Southern Illinois UP, 2000. Print.

Davis, Mike. *City of Quartz: Excavating the Future in Los Angeles*. New York: Verso, 1990. Print.

De Grey, Aubrey. "Life Extension, Human Rights, and the Rational Refinement of Repugnance." *Journal of Medical Ethics* 31 (2005): 659–663. Print.

De Grey, Aubrey, and Michael Rae. *Ending Aging: The Rejuvenation Breakthroughs That Could Reverse Human Aging in Our Lifetime*. New York: St. Martin's, 2007. Print.

De Man, Paul. *Blindness and Insight: Essays in the Rhetoric of Contemporary Criticism*. Minneapolis: U of Minnesota P, 1983. Print.

Deleuze, Gilles. "Appendix: On the Death of Man and Superman." *Foucault*. Trans. Seán Hand. Minneapolis: U of Minnesota P, 1988. 124–132. Print.

———. "Bartleby; or, The Formula." Trans. Daniel W. Smith and Michael A. Greco. *Essays Critical and Clinical*. Minneapolis: U of Minnesota P, 1997. 68–90. Print.

———. "Control and Becoming." *Negotiations, 1972–1990*. Trans. Martin Joughin. New York: Columbia UP, 1995. 169–176. Print.

———. *The Fold: Leibniz and the Baroque*. Trans. Tom Conley. Minneapolis: U of Minnesota P, 1993. Print.

———. "Postscript on Control Societies." *Negotiations*. 177–182. Print.

Deleuze, Gilles, and Felix Guattari. *Anti-Oedipus. Capitalism and Schizophrenia*. Vol. 1. Trans. Robert Hurley, Mark Seem, and Helen R. Lane. Minneapolis: U of Minnesota P, 1983. Print.

———. *A Thousand Plateaus. Capitalism and Schizophrenia*. Vol. 2. Trans. Brian Massumi. Minneapolis: U of Minnesota P, 1987. Print.

———. *What Is Philosophy?* Trans. Hugh Tomlinson and Graham Burchell. New York: Columbia UP, 1994. Print.

Derrida, Jacques. *Archive Fever: A Freudian Impression*. Trans. Eric Prenowitz. Chicago: U of Chicago P, 1995. Print.

———. "Hospitality, Justice, and Responsibility: A Dialogue with Jacques Derrida." *Questioning Ethics: Contemporary Debates in Philosophy*. Ed. Richard Kearney and Mark Dooley. New York: Routledge, 1999. 65–83. Print.

———. "Letter to a Japanese Friend." Trans. David Wood and Andrew Benjamin. *A Derrida Reader: Between the Blinds*. Ed. Peggy Kamuf. New York: Columbia UP, 1991. 269–276. Print.

———. *Of Grammatology*. Corrected Edition. Trans. Gayatri Chakravorty Spivak. Baltimore: Johns Hopkins UP, 1997. Print.

———. *Politics of Friendship*. Trans. George Collins. New York: Verso, 1997. Print.

———. *Spectres of Marx: The State of Debt, the Work of Mourning, and the New International*. Trans. Peggy Kamuf. New York: Routledge, 1994. Print.

Dijksterhuis, E. J. *The Mechanization of the World Picture*. Trans. C. Dikshoorn. London: Clarendon, 1961. Print.

Donald, Merlin. *A Mind So Rare: The Evolution of Human Consciousness*. New York: W. W. Norton, 2001. Print.

Drucker, Peter F. "Knowledge-Worker Productivity: The Biggest Challenge." *California Management Review* 51.2 (Winter 1999): 79–94. Print.

———. *Post-Capitalist Society*. New York: HarperCollins, 1993. Print.

———. *The Unseen Revolution: How Pension Fund Socialism Came to America*. New York: Harper & Row, 1976. Print.

Dyson, George B. *Darwin Among the Machines: The Evolution of Global Intelligence*. Reading, MA: Addison-Wesley, 1997. Print.

Eakin, Emily. "The Latest Theory Is That Theory Doesn't Matter." *New York Times*. Web. 19 April 2003.

Eliot, T. S. *Murder in the Cathedral*. 4th ed. New York: Harcourt, Brace & World, 1935. Print.

Ellerman, A. Denny. *Pricing Carbon: The European Emissions Trading Scheme*. Cambridge: Cambridge UP, 2010. Print.

Elvee, Richard Q. "Introduction." *The End of Science? Attack and Defense*. Nobel Conference XXV. Ed. Elvee. Lanham, MD: 1991. Print.

Ettinger, Robert C. W. *Man Into Superman*. New York: Avon, 1974. Print.

———. *The Prospect of Immortality*. Garden City, NY: Doubleday, 1964. Print.

Finley, M. I. *The Ancient Economy*. Berkeley: U of California P, 1999. Print.

Fisher, Carl Erik, Lisa Chin, and Robert Klitzman. "Defining Neuromarketing: Practices and Professional Challenges." *Harvard Review of Psychiatry* 18.4 (July–August 2010): 230–237. Print.

Flaubert, Gustave. *The Letters of Gustave Flaubert, 1830–1857*. Ed. and Trans. Francis Steegmuller. Cambridge, MA: Belknap/Harvard UP, 1980. Print.

———. *The Letters of Gustave Flaubert, 1857–1880*. Ed. and Trans. Francis Steegmuller. Cambridge, MA: Belknap/Harvard UP, 1982. Print.

Foucault, Michel. "An Aesthetics of Existence." Trans. Alan Sheridan. *Politics, Philosophy, Culture: Interview and Other Writings*. Ed. Lawrence D. Kritzman. 47–53. Print.

———. *The Birth of Biopolitics: Lectures at the Collége de France 1978–1979*. Trans. Graham Burchell. Ed. Michel Senellart. New York: Palgrave Macmillan, 2008. Print.

———. *Discipline and Punish: The Birth of the Prison*. Trans. A. M. Sheridan Smith. New York: Pantheon, 1978. Print.

———. "On the Genealogy of Ethics: An Overview of Work in Progress." *Ethics: Subjectivity and Truth. Essential Works of Foucault 1954–1984*. Vol. 1. Ed. Paul Rabinow. New York: New Press, 1997. 253–280. Print.

———. "The Return of Morality." Trans. Thomas Levin and Isabelle Lorenz. *Politics, Philosophy, Culture: Interviews and Other Writings of Michel Foucault, 1977–1984*. Ed. Lawrence D. Kritzman. New York: Routledge, 1988. 242–255. Print.

———. *Society Must Be Defended: Lectures at the Collége de France, 1975–1976*. Trans. David Macey. Ed. Mauro Bertano and Allessandro Fortana. New York: Picador, 2003. Print.

———. "Technologies of the Self." In *Ethics: Subjectivity and Truth*. 223–251. Print.

———. "Truth and Power." *Power/Knowledge: Selected Interviews and Other Writings, 1972–1977*. Ed. Colin Gordon. Trans. Gordon, Leo Marshall, John Mepham, and Kate Soper. New York: Pantheon, 1980. 109–133. Print.

———. "What Is Enlightenment?" In *Ethics: Subjectivity and Truth*. 303–319. Print.

Frank, Thomas. "Why Johnny Can't Dissent." In *Commodify Your Dissent: Salvos from The Baffler*. Ed. Frank and Matt Weiland. New York: Norton, 1997. 31–45. Print.

Fukuyama, Francis. *Our Posthuman Future: Consequences of the Biotech Revolution*. New York: Picador, 2003. Print.

———. "The World's Most Dangerous Ideas: Transhumanism." *Foreign Policy* 144 (September–October 2004): 42–43. Print.

Galloway, Alexander R. *Protocol: How Control Exists after Decentralization*. Cambridge, MA: MIT P, 2004. Print.

Gigerenzer, Gerd, et al. *The Empire of Chance: How Probability Changed Science and Everyday Life*. Cambridge: Cambridge UP, 1989. Print.

Gilroy, Paul. *The Black Atlantic: Modernity and Double Consciousness*. Cambridge, MA: Harvard UP, 1993. Print.

Ginsburg, Geoffrey S., and Jeanette J. McCarthy. "Personalized Medicine: Revolutionizing Drug Discovery and Patient Care." *Trends in Biotechnology* 19.12 (December 2001): 491–496. Print.

Grossberg, Lawrence. *We Gotta Get Out of This Place: Popular Conservatism and Postmodern Culture.* New York: Routledge, 1992. Print.

Guattari, Félix. "Transversality." Trans. Rosemary Sheed. *Molecular Revolution: Psychiatry and Politics.* New York: Penguin, 1984. 11–18. Print.

Guillén, Mauro F. *The Taylorized Beauty of the Mechanical: Scientific Management and the Rise of Modernist Architecture.* Princeton, NJ: Princeton UP, 2006. Print.

Habermas, Jürgen. *The Future of Human Nature.* Trans. William Rehg, Max Pensky, and Hella Beister. Cambridge: Polity, 2003. Print.

———. *The Structural Transformation of the Public Sphere: An Inquiry Into a Category of Bourgeois Society.* Trans. Thomas Burger with Frederick Lawrence. Cambridge, MA: MIT P, 1989. Print.

Hacking, Ian. *The Taming of Chance.* Cambridge: Cambridge, UP: 1990. Print.

Halloran, S. Michael. "Aristotle's Concept of Ethos, or If Not His Somebody Else's." *Rhetoric Review* 1.1 (September 1982): 58–63. Print.

Han, Seunghee, Jennifer S. Lerner, and Richard Zeckhauser. "The Disgust-Promotes-Disposal Effect." *Journal of Risk and Uncertainty* 44 (2012): 101–113. Print.

Haraway, Donna J. "A Cyborg Manifesto: Science, Technology, and Socialist-Feminism in the Late Twentieth Century." *Simians, Cyborgs, and Women: The Reinvention of Nature.* New York: Routledge, 1991. 149–182. Print.

Hardt, Michael. "Affective Labor." *boundary 2* 26.2 (1999): 89–100. Print.

———. "The Withering of Civil Society." *Deleuze & Guattari: New Mappings in Politics, Philosophy, and Culture.* Ed. Eleanor Kaufman and Kevin Jon Heller. Minneapolis: U of Minnesota P, 1988. 23–39. Print.

Hardt, Michael, and Antonio Negri. *Empire.* Cambridge, MA: Harvard UP, 2000. Print.

———. *Multitude: War and Democracy in the Age of Empire.* New York: Penguin, 2004. Print.

Harris, Oliver. "Cut-Up Closure: The Return to Narrative." *William S. Burroughs at the Front: Critical Reception, 1959–1989.* Ed. Jennie Skerl and Robin Lyndenberg. Carbondale: Soutern Illinois UP, 1991. 251–262. Print.

Havelock, Eric A. *Preface to Plato.* Cambridge, MA: Belknap UP, 1963. Print.

Hawhee, Debra. "Burke and Nietzsche." *Quarterly Journal of Speech* 85 (1999): 129–145. Print.

Hayles, N. Katherine. *How We Became Posthuman: Virtual Bodies in Cybernetics, Literature, and Informatics.* Chicago: U of Chicago P, 1999. Print.

———. "Wrestling with Transhumanism." *The Global Spiral: A Publication of Metanexus Institute* (June 2008). Web. 16 March 2010.

Hegel, Georg Wilhelm Friedrich. *Introduction to the Philosophy of History: With Selections from the Philosophy of Right.* Trans. Leo Rauch. Indianapolis, IN: Hackett, 1988. Print.

———. *Lectures on the History of Philosophy.* Vol 1. *Greek Philosophy to Plato.* Trans. E. S. Haldane. Lincoln: U of Nebraska P, 1995. Print.

Heidegger, Martin. "The Age of the World Picture." Trans. William Lovitt. *The Question Concerning Technology and Other Essays.* New York: Harper & Row, 1977. 115–151. Print.

———. *Being and Time.* Trans. John Macquarrie and Edward Robinson. San Francisco: Harper & Row, 1962. Print.

———. *Discourse On Thinking.* Trans. John M. Anderson and E. Hans Freund. New York: Harper & Row, 1966. Print.

———. "The End of Philosophy and the Task of Thinking." *On Time and Being.* Trans. Joan Stambaugh. Chicago: U of Chicago P, 1972. 55–73. Print.

———. *The History of the Concept of Time.* Trans. Theodore Kisiel. Bloomington: Indiana UP, 1985. Print.

———. *Introduction to Metaphysics.* Trans. Gregory Fried and Richard Polt. New Haven, CT: Yale UP, 2000. Print.

———. "Only a God Can Save Us Now: An Interview with Martin Heidegger." Trans. M. P. Alter and J. D. Caputo. *Philosophy Today* 20 (1876): 267–284. Print.

———. "The Origin of the Work of Art." Trans. Albert Hofstadter. *Basic Writings from* Being and Time *(1927) to* The Task of Thinking *(1964).* Ed. David Farrell Krell. New York: HarperCollins, 2008. 139–213. Print.

———. "Overcoming Metaphysics." *The End of Philosophy.* Trans. Joan Stambaugh. Chicago: U of Chicago P, 1973. 84–110. Print.

———. "What Are Poets For?" *Poetry, Language, Thought.* Trans. Albert Hofstadter. New York: Harper & Rowe, 1971. 91–142. Print.

Heims, Steve Joshua. *Constructing a Social Science for Postwar America: The Cybernetics Group, 1946–1953.* Cambridge, MA: MIT P, 1993. Print.

Hemmings, Clare. "Invoking Affect: Cultural Theory and the Ontological Turn." *Cultural Studies* 19.5 (2005): 548–567. Print.

Henderson, Mark. "Cancer-Free 'Designer Babies' Get Approval." *Times Online* (London). Web. 1 Nov. 2004.

Holland, Eugene. "From Schizophrenia to Social Control." *Deleuze & Guattari: New Mappings in Politics, Philosophy, and Culture.* Ed. Eleanor Kaufman and Kevin Jon Heller. Minneapolis: U of Minnesota P, 1988. 65–73. Print.

Honig, Bonnie. "Antigone's Two Laws: Greek Tragedy and the Politics of Humanism." *New Literary History* 41.1 (Winter 2010): 1–33. Print.

Horgan, John. *The End of Science: Facing the Limits of Knowledge in the Twilight of the Scientific Age.* New York: Broadway, 1997. Print.

Horkheimer, Max. "Traditional and Critical Theory." Trans. Matthew J. O'Connell. *Critical Theory: Selected Essays.* New York: Continuum, 1982.188–243. Print.

Horkheimer, Max, and Theodor W. Adorno. *Dialectic of Enlightenment: Philosophical Fragments.* Trans. Edmund Jephcott. Ed. Gunzelin Schmid Noerr. Stanford, CA: Stanford UP, 2002. Print.

Huxley, Julian. "Transhumanism." *New Bottles for Old Wine.* London: Chatto & Windus, 1959. 13–17. Print.

Jameson, Fredric. "Culture and Finance Capital." *The Cultural Turn: Selected Writings on the Postmodern, 1983–1998.* New York: Verso, 1998. 136–161. Print.

———. *Postmodernism, or, The Cultural Logic of Late Capitalism.* Durham, NC: Duke University Press, 1991. Print.

Jay, Martin. *Downcast Eyes: The Denigration of Vision in Twentieth-Century Thought.* Berkeley: U of California P, 1993. Print.

Jenniges, Amy. "Mild Kingdom: Two Seattle Dudes Play Homeless, Upload Their Urban Safari for World to See." *The Stranger* (Seattle). Web. 30 July 2002.

Johnston, Adrian. *Badiou, Zizek, and Political Transformations: The Cadence of Change.* Evanston, IL: Northwestern UP, 2009. Print.

Kant, Immanuel. *Anthropology from a Pragmatic Point of View.* Trans. and Ed. Robert B. Louden. Cambridge: Cambridge UP, 2006. Print.

———. *Critique of Pure Reason.* Trans. Marcus Weigelt. New York: Penguin, 2007. Print.

Kass, Leon R. "L'Chaim and Its Limits: Why Not Immortality?" *The Fountain of Youth: Cultural, Scientific, and Ethical Perspectives on a Biomedical Goal.* Ed.

Stephen G. Post and Robert H. Binstock. Oxford: Oxford UP, 2004. 304–320. Print.

Keller, Evelyn Fox. "Models of and Models For: Theory and Practice in Contemporary Biology." *Philosophy of Science* 67 (2000): S72–S86. Print.

Kelley, Donald R. "Philodoxy: Mere Opinion and the Question of History." *Journal of the History of Philosophy* 34.1 (January 1996): 117–132. Print.

Kennedy, George. Trans. "Gorgias." *The Older Sophists: A Complete Translation of the Fragments in* Die Fragmente der Vorsokratiker, *Edited by Diels-Kranz. With a New Edition of* Antiphon *and of* Euthydemus. Ed. Rosamund Kent Sprague. Indianapolis, IN: Hackett, 2001. Print.

Knickerbocker, Conrad. "White Junk." *Burroughs Live: The Collected Interviews of William S. Burroughs, 1960–1997.* Ed. Sylvère Lotringer. Los Angeles: Semiotext(e), 2001. 60–81. Print.

Koepp, M. J., et al. "Evidence for Striatal Dopamine Release During a Video Game." *Nature* 393 (1998): 266–268. Print.

Kowsh, Kate. "Vyuz Talks to Bumfights Creator Ryen McPherson." Vyuz.com. Web. 20 Feb. 2006.

Kurke, Leslie. *Coins, Bodies, Games, and Gold: The Politics of Meaning in Archaic Greece.* Princeton, NJ: Princeton UP, 1999. Print.

Kurzweil, Ray. *The Singularity Is Near: When Humans Transcend Biology.* New York: Viking, 2005. Print.

Lafontaine, Céline. "The Cybernetic Matrix of 'French Theory.'" *Theory, Culture, & Society* 24.5 (2007): 27–46. Print.

Latour, Bruno. "Why Has Critique Run Out of Steam? From Matters of Fact to Matters of Concern." *Critical Inquiry* 30 (Winter 2004): 225–248. Print.

Lazzarato, Maurizio. "Immaterial Labor." Trans. Paul Colilli and Ed Emory. *Radical Thought in Italy: A Potential Politics.* Ed. Palo Virno and Michael Hardt. Minneapolis: U of Minnesota P, 1996. 133–157. Print.

Le Goff, Jacques. *Time, Work, and Culture in the Middle Ages.* Trans. Arthur Goldhammer. Chicago: U of Chicago P, 1980. Trans.

Lerner, Jennifer S., Deborah A. Small, and George Loewenstein. "Heart Strings and Purse Strings: Carryover Effects of Emotions on Economic Decisions." *Psychological Science* 15.5 (2004): 337–341. Print.

Levinas, Emmanuel. *Totality and Infinity: An Essay on Exteriority.* Trans. Alphonso Lingis. Pittsburgh, PA: Duquesne UP, 1969. Print.

Levy, Jonathan Ira. "Contemplating Delivery: Futures Trading and the Problem of Commodity Exchange in the United States, 1875–1905." *The American Historical Review* 111.2 (April 2006): 307–335. Print.

Levy, Steven. "In Every Voter, a 'Microtarget.'" *Washington Post* 23 April 2008. D01. Print.

Lewis, David, and Darren Bridger. *The Soul of the New Consumer: Authenticity— What We Buy and Why in the New Economy.* London: Nicholas Brealey, 2000. Print.

Leys, Ruth. "The Turn to Affect: A Critique." *Critical Inquiry* 37.3 (Spring 2001): 434–472. Print.

Lindstrom, Martin. *Buyology: Truth and Lies About Why We Buy.* New York: Doubleday, 2008. Print.

Lovelock, James. *Gaia: A New Look at Life on Earth.* Oxford: Oxford UP, 2000. Print.

Lukacs, Georg. "Thoughts toward an Aesthetic of the Cinema." Trans. Janelle Blankenship. *Polygraph* 13 (2001): 13–18. Print.

Lynn, Greg. *Animate Form.* New York: Princeton Architectural Press, 1998. Print.

Lyotard, Jean-Francois. "Appendix: Answering the Question: What Is Postmodernism?" *The Postmodern Condition: A Report on Knowledge.* Trans. Geoff Bennington and Brian Massumi. Minneapolis: U of Minnesota P, 1984. 70–84. Print.

———. "Return Upon the Return." *Toward the Postmodern.* Ed. Robert Harvery and Mark S. Roberts. Atlantic Highlands, NJ: Humanities Press, 1993. 192–206. Print.

MacKenzie, Donald, and Yuval Millo. "Constructing a Market, Performing Theory: The Historical Sociology of a Financial Derivatives Exchange." *American Journal of Sociology* 109.1 (July 2003): 107–145. Print.

Mailloux, Steven. "One Size Doesn't Fit All: The Contingent Universality of Rhetoric." *Sizing Up Rhetoric.* Ed. David Zarefsky and Elizabeth Benacka. Long Grove, IL: Waveland, 2008. 7–19. Print.

Manovich, *The Language of New Media.* Cambridge, MA: MIT P, 2001. Print.

Marazzi, Christian. *Capital and Language: From the New Economy to the War Economy.* Trans. Gregory Conti. Los Angeles: Semiotext(e), 2008. Print.

Marcus, George E., W. Russell Neuman, and Michael MacKuen. *Affective Intelligence and Politcal Judgment.* Chicago: U of Chicago P, 2000. Print.

Margulis, Lynn, and Dorian Sagan. *Acquiring Genomes: A Theory of the Origins of Species.* Oxford: Oxford UP, 2000. Print.

Marx, Karl. *Capital.* Vol. 1. Ed. Friedrich Engels. Trans. Ben Fowkes. New York: Vintage, 1977. Print.

———. *Grundrisse: Foundations of the Critique of Political Economy.* Trans. Martin Nicolaus. London: Penguin, 1973. Print.

Marx, Karl and Friedrich Engels. *The German Ideology Parts I & III.* Ed. R. Pascal. Mansfield Centre, CT: Martino, 2011. Print.

Massumi, Brian. "The Autonomy of Affect." *Cultural Critique* 31 (Fall 1995): 83–109. Print.

———. "Navigating Movements." *Hope: New Philosophies for Change.* Ed. Mary Zournazi. London: Lawrence & Wishart, 2002. 210–242. Print.

———. *Parables for the Virtual: Movement, Affect, Sensation.* Durham, NC: Duke UP, 2002. Print.

McComiskey, Bruce. *Gorgias and the New Sophistic Rhetoric.* Carbondale: Southern Illinois UP, 2002. Print.

McKirahan, Richard D. *Philosophy Before Socrates: An Introduction with Texts and Commentary.* Indianapolis, IN: Hackett, 1994. Print.

Mead, Rebecca. "Rage Machine: Andre Breitbart's Empire of Bluster." *New Yorker* 24 May 2010. Web. 14 March 2011.

Meek, James. "Why You Are First in the Great Gene Race." *The Guardian* (London) 15 November 2000. Web. 12 March 2010.

Miller, Stanley A. "Death of a Game Addict" *Milwaukee Journal Sentinel* 31 March 2002. Web. 12 April 2004.

Mitchell, W. J. T. "Medium Theory: Preface to the 2003 *Critical Inquiry* Symposium." *Critical Inquiry* 30.2 (Winter 2004): 324–335. Print.

Mitchell, William J. *City of Bits: Space, Place, and Infobahn.* Cambridge, MA: MIT P, 1995. Print.

Moulier-Boutang, Yann. *Cognitive Capitalism.* Cambridge: Polity, 2012. Print.

Muckelbauer, John. *The Future of Invention: Rhetoric, Postmodernism, and the Problem of Change.* Albany: State U of New York P, 2008. Print.

Mulherin, J. Harold, Jeffrey M. Netter, and James A. Overdahl. "Prices Are Property: The Organization of Financial Exchanges from a Transaction Cost Perspective." *Journal of Law & Economics* 59 (October 1991): 591–644. Print.

Naas, Michael. *Turning: From Persuasion to Philosophy.* Atlantic Highlands, NJ: Humanities Press, 1994. Print.

National Science Board. "A Companion to Science and Engineering Indicators 2008." nsf.gov. Web. 20 April 2012.

Nealon, Jeffrey T. *Alterity Politics: Ethics and Performative Subjectivity.* Durham, NC: Duke UP, 1998. Print.

———. *Foucault Beyond Foucault: Power and Its Intensifications Since 1984.* Stanford, CA: Stanford UP, 2008. Print.

Negri, Antonio. "Value and Affect." Trans. Michael Hardt. *boundary 2* 26.2 (1999): 77–87. Print.

Nietzsche, Friedrich. *The Birth of Tragedy.* Trans. Ronald Speirs. Ed. Raymond Geuss and Ronald Speirs. Cambridge: Cambridge UP, 1999. Print.

———. *Human, All Too Human.* Trans. R. J. Hollingdale. Cambridge: Cambridge UP, 1996. Print.

———. "Homer's Contest." *On the Genealogy of Morality.* 174–181. Print.

———. *On the Genealogy of Morality.* Rev. Student Edition. Trans. Carol Deithe. Ed. Keith Ansell-Pearson. Cambridge: Cambridge UP, 2007. Print.

———. *Twilight of the Idols or How to Philosophize with a Hammer.* Trans. Judith Norman. *The Anti-Christ, Ecce Homo, Twilight of the Idols, and Other Writings.* Ed. Aaron Ridley and Judith Norman. Cambridge: Cambridge UP, 2005. 153–229. Print.

Nightingale, Andrea Wilson. *Spectacles of Truth in Classical Greek Philosophy: Theoria in Its Cultural Context.* Cambridge: Cambridge UP, 2004. Print.

North, John. *God's Clockmaker: Richard of Wallingford and the Invention of Time.* New York: Hambledon and London, 2005. Print.

Pignarre, Phillipe, and Isabelle Stengers. *Capitalist Sorcery: Breaking the Spell.* Trans. and Ed. Adrew Goffey. New York: Palgave MacMillan, 2011. Print.

Penn, Mark J., with E. Kinney Zalesne, *Microtrends: The Small Forces Behind Tomorrow's Big Changes.* New York: Hachette, 2007. Print.

Plato. *Apology.* Trans. G. M. A. Grube. *Complete Works.* Ed. John M. Cooper. Indianapolis, IN: Hackett, 1997. 17–36. Print.

———. *Gorgias.* Trans. Donald J. Zeyl. In *Complete Works.* 791–869. Print.

———. *Laws.* Trans. Trevor J. Saunders. In *Complete Works.* 1318–1616. Print.

———. *Meno.* Trans. G. M. A. Grube. In *Complete Works.* 870–897. Print.

———. *Philebus.* Trans. Dorothea Frede. In *Complete Works.* 398–456. Print.

———. *Protagoras.* Trans. Stanley Lombardo and Karen Bell. In *Complete Works.* 746–790. Print.

———. *The Republic.* Trans. G. M. A. Grube and C. D. C. Reeve. In *Complete Works.* 971–1223. Print.

Pope, Alexander. "An Essay on Man." *Selected Poetry and Prose.* Ed. William K. Wimsatt. 2nd ed. New York: Holt, Rinehart, and Winston, 1972. 191–232. Print.

Rasch, William. *Niklas Luhmann's Modernity: The Paradoxes of Differentiation.* Stanford, CA: Stanford UP, 2000. Print.

Rohde, Erwin. *Psyche: The Cult of Souls and Belief in Immortality among the Greeks.* 2 vols. Trans. W. B. Willis. Eugene, OR: Wipf & Stock, 2006. Print.

Rosenblueth, Arturo, and Norbert Wiener. "Purposeful and Non-Purposeful Behavior." *Philosophy of Science* 17.4 (October 1950): 318–326. Print.

Rueckert, William. *Kenneth Burke and the Drama of Human Relations.* 2nd ed. Berkeley: U of California P, 1982. Print.

Sasaki, Mutsoro. "Morphogenesis of Flux Structure." *From Control to Design: Parametric/Algorithmic Architecture.* Ed. Tomoko Sakamoto and Albert Ferre. Barcelona, Spain: Actar-D, 2007. 68–89. Print.

Schaps, David M. "Socrates and the Socratics: When Wealth Becomes a Problem." *The Classical World* 96.2 (Winter 2003): 131–157. Print.

Schiappa, Edward. "Second Thoughts on the Critiques of Big Rhetoric." *Philosophy and Rhetoric* 34.3 (2001): 260–274. Print.

Schnur, Dan. "The Affect Effect in the Very Real World of Poltical Campaigns." *The Affect Effect: Dynamics of Emotion in Political Thinking and Behavior.* Ed. W. Russell Neuman, George E. Marcus, Ann N. Crigler, and Michael MacKuen. Chicago: U of Chicago P, 2007. 357–374. Print.

Seaford, Richard. *Money and the Early Greek Mind: Homer, Philosophy, Tragedy.* Cambridge: Cambridge UP, 2004. Print.

Sedgwick, Eve Kosofsky. *Touching Feeling: Affect, Pedagogy, Performativity.* Durham, NC: Duke UP, 2003. Print.

Sedgwick, Eve Kosofsky, and Adam Frank. "Shame in the Cybernetic Fold: Reading Silvan Tomkins." *Critical Inquiry* 21.2 (Winter 1995): 496–522. Print.

Shell, Marc. *Money, Language, and Thought: Literary and Philosophical Economies from the Medieval to the Modern Era.* Berkeley: U of California P, 1982. Print.

Shyrock, Richard H. "The History of Quantification in Medical Science." *Isis* 52.2 (June 1961): 215–237. Print.

Simmel, Georg. *The Philosophy of Money.* 3rd ed. Trans. Tom Bottomore and David Frisby. Ed. David Frisby. New York: Routledge, 2004. Print.

Simon, Herbert. "Motivational and Emotional Controls of Cognition." *Psychological Review* 74.1 (1967): 29–39. Print.

Steiner, Deborah T. "Herodotus and the Language of Metals." *Helios* 22 (1995): 36–64. Print.

Stengers, Isabelle, and Didier Gille. "Time and Representation." *Power and Invention: Situating Science.* Minneapolis: U of Minnesota P, 1997. 177–214. Print.

Strathausen, Carsten. "A Critique of Neo-Left Ontology." *Postmodern Culture* 16.3 (2006). Web. 18 March 2011.

Svoboda, Elizabeth. "All Politics Is Microtargeting: Six Political Strategists Who Study What You Eat, What You Drive, and Where You Shop." *Fast Company.* Web. 10 April 2009.

Sweet, Ken. "NYSE's Grab at a 3.7 Quadrillion Market." *CNN Money.* Web. 14 Feb. 2011.

Swijtink, Zeno G. "The Objectification of Observation: Measurement and Statistical Methods in the Nineteenth Century." *The Probabilistic Revolution.* Vol. 1, *Ideas in History.* Ed. Lorenz Kruger, Lorraine J. Daston, and Michael Heidelberger. Cambridge, MA: MIT P, 1987. 260–285. Print.

Tapscott, Don, and Anthony D. Williams. *Wikinomics: How Mass Collaboration Changes Everything.* New York: Portfolio, 2006. Print.

Tipler, Frank J. *The Physics of Immortality: Modern Cosmology, God and the Resurrection of the Dead.* New York: Anchor, 1995. Print.

Thom, René. *Structural Stability and Morphogenesis.* Trans. D. H. Fowler. Reading, MA: W. A. Benjamin, 1975. Print.

Thomson, Iain D. *Heidegger on Ontotheology: Technology and the Politics of Education.* Cambridge: Cambridge UP, 2005. Print.

Tomkins, Sylvan S., and Robert McCarter. "What and Where Are the Primary Affects? Some Evidence for a Theory." *Exploring Affect: The Selected Writings of Silvan S. Tomkins.* Ed. E. Virginia Demos. Cambridge: Cambridge UP, 1995. Print.

Tronti, Mario. "The Strategy of Refusal." *Working Class Autonomy and the Crisis.* Ed. Red Notes Collective. London: Red Notes, 1979. 7–21. Print.

Turing, A. M. "Computing Machinery and Human Intelligence." *Mind: A Quarterly Review of Psychology and Philosophy* 59.236 (1950): 433–460. Print.

———. "Solvable and Unsolvable Problems." *Science News* 31 (1954): 7–23. Print.

Vasilou, Iakovos. *Aiming at Virtue in Plato.* Cambridge: Cambridge UP, 2008. Print.

Veblen, Thorstein. *The Theory of the Leisure Class: An Economic Study of Institutions.* New York: Modern Library, 1931. Print.

Vesely, Dalibor. *Architecture in the Age of Divided Representation: The Question of Creativity in the Shadow of Production.* Cambridge, MA: MIT P, 2004. Print.

Virno, Paolo. *A Grammar of the Multitude: For an Analysis of Contemporary Forms of Life.* Trans. Isabella Bertoletti, James Cascaito, and Andrea Casson. Los Angeles: Semiotext(e), 2004. Print.

———. "The Ambivalence of Disenchantment." Trans. Michael Turits. In *Radical Thought in Italy: A Potential Politics.* Ed. Palo Virno and Michael Hardt. Minneapolis: U of Minnesota P, 1996. 13–34. Print.

Vitanza, Victor J. *Negation, Subjectivity, and the History of Rhetoric.* Albany: State U of New York P, 1997. Print.

Watanabe, Satosi. "Epistemological Implications of Cybernetics." *Proceedings of the Fourteenth World Congress of Philosophy.* Vienna: Herder, 1968. 594–600. Print.

Weiss, Rick. "Of Mice, Men and in-Between: Scientist Debate Blending of Human, Animal Forms." *Washington Post.* Web. 20 Nov. 2004.

Weizenbaum, Joseph. *Computing Power and Human Reason: From Judgment to Calculation.* San Francisco: W. H. Freeman and Company, 1976. Print.

Welch, Kathleen E. *Electric Rhetoric: Classical Rhetoric, Oralism, and a New Literacy.* Cambridge, MA: MIT P, 1999. Print.

Wiener, Norbert. *Cybernetics; or, Control and Communication in the Animal and Machine.* 2nd ed. Cambridge, MA: MIT P, 1961. Print.

———. *The Human Use of Human Beings: Cybernetics and Society.* New York: De Capo, 1988. Print.

———. *I Am a Mathematician: The Later Life of a Prodigy.* Cambridge, MA: MIT P, 1964. Print.

Wess, Robert. *Kenneth Burke: Rhetoric, Subjectivity, Postmodernism.* Cambridge: Cambridge UP, 1996. Print.

Williams, Raymond. "Advertising: the Magic System." *Culture and Materialism.* New York: Verso, 1980. 170–195. Print.

Wilson, Elisabeth A. *Neural Geographies: Feminism and the Microstructure of Cognition.* New York: Routledge, 1998. Print.

Wöfflin, Heinrich. *Renaissance and Baroque.* Trans. Kathryn Simon. Ithaca, NY: Cornell UP, 1966. Print.

Wolfe, Cary. *Animal Rites: American Culture, The Discourse of Species, and Posthumanist Theory.* Chicago: U of Chicago P, 2003. Print.

———. *What Is Posthumanism?* Minneapolis: U of Minnesota P, 2010. Print.

Worsham, Lynn. "Going Postal: Pedagogic Violence and the Schooling of Emotion." *JAC* 18.2 (1998): 213–245. Print.

———. "Coming to Terms: Theory, Writing, Politics." *Rhetoric and Composition as Intellectual Work.* Ed. Gary A. Olsen. Carbondale: Southern Illinois UP, 2002. 101–114. Print.

Worth, Jess. "Punk Rock Capitalism?" *New Internationalist* 395 (Nov 2006). Web 15 March 2010.

Wright, Erik Olin. *Envisioning Real Utopias.* New York: Verso, 2010. Print.

Zizek, Slajov. *The Fragile Absolute, or, Why Is the Christian Legacy Worth Fighting For?* New York: Verso, 2008. Print.

————. "Notes toward a Politics of Bartleby: The Ignorance of Chicken." *Comparative American Studies* 4.4 (2006): 375–394. Print.

————. *Organs without Bodies: On Deleuze and Consequences.* New York: Routledge, 2004. Print.

————. *The Parallax View.* Cambridge, MA: MIT P, 2006. Print.

————. *The Sublime Object of Ideology.* New York: Verso, 1989. Print.

————. *Tarrying with the Negative: Kant, Hegel, and the Critique of Ideology.* Durham, NC: Duke UP, 1993. Print.

————. *The Ticklish Subject: The Absent Centre of Political Ontology.* New York: 2000. Print.

Index

For Product Safety Concerns and Information please contact our EU
representative GPSR@taylorandfrancis.com
Taylor & Francis Verlag GmbH, Kaufingerstraße 24, 80331 München, Germany

www.ingramcontent.com/pod-product-compliance
Lightning Source LLC
Chambersburg PA
CBHW071425050326
40689CB00010B/1985